Polish-Japanese symposium on Hemostasis and Circulation   August 8 and 9, 1991   Grand Hotel, Hamamatsu

A. Takada · A.Z. Budzynski (Eds.)

Hemostasis
and Circulation

With 110 Illustrations

Springer Japan KK

A. TAKADA
Department of Physiology, Hamamatsu University, School of Medicine, Hamamatsu, Shizuoka,
431-31 Japan

A.Z. BUDZYNSKI
Department of Biochemistry, Temple University, School of Medicine, Philadelphia, USA

ISBN 978-4-431-70096-8   ISBN 978-4-431-66925-8 (eBook)
DOI 10.1007/978-4-431-66925-8

© Springer Japan 1992
Originally published by Springer-Verlag Tokyo in 1992.

PREFACE

The First Polish-Japanese Symposium on Hemostasis and Circulation was held on August 8 and 9, 1991 at Hamamatsu Grand Hotel, Hamamatsu, Shizuoka-ken, Japan.   The meeting was officially supported by the Japanese Ministry of Education, Science and Culture.   The financial support contributed by many industries and pharmaceutical companies made this meeting possible.   The Organizing Committee of the Symposium was formed at Hamamatsu University School of Medicine under the chairmanship of Professor Shukichi Sakaguchi with Professor Akikazu Takada as a secretary.

The First Polish-Japanese Symposium on Hemostasis and Circulation brought together scientists from three countries and provided a forum for promotion of scientific cooperation.   The topics addressed at the Symposium are of interest and importance not only for participants but also for investigators throughout the world.   A refreshing approach has been accomplished by joining problems of hemostasis with those of the circulatory blood, an interrelationship addressing pathophysiological circumstances.

The Polish investigators have a long tradition in pursuing problems of hemostasis and thrombosis.   Initially, Dr. H. Kowarzyk and then Drs. Buluk and E. Kowalski inspired and educated many investigators who working in Poland and abroad achieved international recognition.   The recent political and economic changes in Poland are harbingers of progress and stabilization that will provide a better foundation for international scientific cooperation.   A good example are several Polish investigators who came to Hamamatsu and participated in the first step in this direction.

Finally, we would like to express our sincere thanks to members of the Department of Physiology of Hamamatsu University, especially Dr. Yumiko Takada, who helped to accomplish the successful symposium.

August 21, 1991

Akikazu Takada
Andrei Z. Budzynski
Hamamatsu, Japan

OVERVIEW

Andrei Z. Budzynski and Akikazu Takada

A variety of topics presented at the Symposium in 34 talks addressed current issues of hemostasis and circulation. These have been grouped into problems of interaction with thrombin, modulation of coagulation, platelet function, fibrinolysis, topics pertinent to circulation, and clinical aspects of hemostasis. Selected reports have been highlighted before to illustrate the scope of the Symposium.

Thrombin catalyzes the conversion of fibrinogen to fibrin. The structure of a fibrin polymerization site in the gamma chain, that directs the formation of a clot, has been delineated after chemical crosslinking of a complementary peptide (Cierniewski and Budzynski). Specific degradation of fibrinogen in the blood was demonstrated to be the mechanism of blood anticoagulation by hementin, an enzyme occurring in the salivary gland of the giant South American leech (Budzynski). The cellular mechanism of plasma thrombomodulin regulation, involving among others cytokines and glucocorticoids, affects its level and may be of pathologic significance (Ohdama et al). Platelet functions are modulated by a variety of agents isolated from snake venoms, in particular by disintegrins, alpha-fibrinogenases, and thrombin-like enzymes (Teng and Huang). In an electron microscope study the storage of GPIIb/IIIa in the alpha-granules and its migration and attachment to the cell surface skeleton, upon activation of platelets have been shown (Suzuki). Platelet aggregation occurs not only upon the action of agonists, but also as a result of shear stress at low or high velocities of flow (Ikeda and Handa). As ethanol inhibits ADP-induced platelet aggregation, it has been shown that its effect is mediated through serotonergic mechanism which is not associated with the serotonin platelet receptor (Buczko). Regarding regulation of fibrinolysis, it was demonstrated that a peptide from the amino-terminal domain of Glu-plasminogen, containing -Lys-Glu- sequence, is important for the lysine binding site and enhances plasminogen activation (Urano et al). Stimulated neutrophils secrete proteolytic enzymes affecting platelet aggregation and clot lysis (Kopec). For example, single chain urokinase is inactivated by neutrophil elastase (Kanayama et al); however, cathepsin B activates the soluble and cell receptor-bound form of the proenzyme (Kobayashi et al). Extracorporeal circulation in oxygenators, artificial dialysis devices, and during plasmapheresis exposes blood to thrombogenic surfaces; blood response is complex involving contact activation of coagulation, complement activation, and cell adhesion (Matsuda). A novel direction was attempts to assess the influence of the central nervous system on coagulation and fibrinolysis. For instance, induction of a mental stress resulted in increase of catecholamines and decrease of plasminogen activator inhibitor that led to activation of both platelets and fibrinolysis (Takada et al).

Contents

PLATELETS

FIBRINOLYSIS

CIRCULATION

CLINICAL STUDIES

Basic and
Functional Studies

PRIMARY POLYMERIZATION SITES IN THE D-DOMAIN OF HUMAN FIBRINOGEN

C.S. CIERNIEWSKI[1] and A.Z. BUDZYNSKI[2]

[1] Department of Biophysics, Medical School in Lodz, Poland,
[2] Department of Biochemistry, Temple University School of Medicine,
  Philadelphia, PA 19140

INTRODUCTION

The conversion of fibrinogen into a fibrin clot is initiated by the limited thrombin proteolysis resulting in the release of fibrinopeptide A and the exposure of NH_2-terminal polymerization sites "A" in the E domain of the fibrinogen molecule.[1] Complementary polymerization sites "a" capable of interacting with the NH_2-terminal sites have been attributed to the D domain, but their exact location in the fibrinogen molecule is uncertain.[2-5] There are inconsistent preliminary data indicating that sites "a" may be present in sequences either γ_{95-264}[4] or $\gamma_{265-411}$.[5] The present studies were initiated to assess this discrepancy and identify the location of a polymerization sites "a" in the D domain.

MATERIALS AND METHODS

Peptides - A 9-amino acid fibrinogen peptide fragment (GPRVVERHK) corresponding to residues 17-24 of the Aα chain, containing Lys at the COOH terminus which is not present in the natural sequence, was synthesized by Peninsula. The peptide was dissolved in 0.15 M NaCl, 0.01 M sodium phosphate, pH 7.3 (PBS) and radioiodinated using IODO-beads (Bio-Rad).

Inhibition of fibrin monomer polymerization by GPRVVERHK was measured spectrophotometrically at 350 nm.[6]

Peptide cross-linking - Fibrinogen dissolved in PBS was mixed with ^{125}I-GPRVVERHK and binding proceeded for 30 min at room temperature. The primary cross-linking agent used in this study was disuccinimidyl suberate (DSS, Pierce Chemical Co.). The cross-linking reactions were terminated after 20 min at room temperature by addition of 1 M glycine buffer, pH. 8.0. Fibrinogen cross-linked with the radioiodinated GPRVVERHK was hydrolysed by plasmin (0.5 CTA) for 2 hours at 37° C in the presence (10 mM) or absence of calcium ions. Digestion was stopped by addition of trasylol (500 KIU), aliquots of the digest were mixed with Laemmli Sample buffer and separated by SDS PAGE.[7]

Preparative recovery of the cross-linked GPRVVERHK - Fragment D_1, the major fibrinogen cleavage product cross-linked with ^{125}I-GPRVVERHK, was separated from plasmic digest of fibrinogen by immunoaffinity chromatography on anti-FgD antibodies immobilized on Sepharose 4B. FgD$_1$-GPRVVERHK was further digested with 0.1 CTA units of plasmin in the presence of 20 mM EGTA for 24 hours at 37° C. The digest was separated by HPLC on the reverse phase C_{18} column. Peptide fractions containing the radioactive GPRVVERHK were rechromatographed and directly applied into a gas phase sequenator (Applied Biosystem model 475A).

RESULTS

The peptide GPRVVERHK bound to fibrin monomers and when added in a 4000-fold molar excess abolished their polymerization. However, it was approximately 10-fold less potent in inhibiting of fibrin monomer polymerization than GPRP. IC_{50} of GPRVVERHK and GPRP were 385 µM and 39 µM, respectively.

To locate a binding site for GPRVVERHK in the fibrinogen molecule, this protein was preincubated with 22.59 µM of ^{125}I-GPRVVERHK for 30 minutes at room temperature and then chemical cross-reacting reagent DSS was added. After 20 minutes incubation at room temperature, the DSS was neutralized with 1 M glycine buffer, and samples of fibrinogen solution were analyzed by SDS-PAGE under reducing conditions. Autoradiograms indicated that the radioactive peptide was predominantly bound to the γ chain (Fig. 1:curve 1). Though the concentration of the DSS was very low (0.2 mM), there was a residual cross-linking of the polypeptide chains within the fibrinogen molecule. The specificity of the cross-linking of ^{125}I-GPRVVERHK was explored by assessing the capacity of nonlabeled GPRVVERHK to inhibit the interaction of the radioiodinated peptide with polypeptide chains of fibrinogen. To estimate the relative inhibition of the cross-linking reaction by different concentrations of

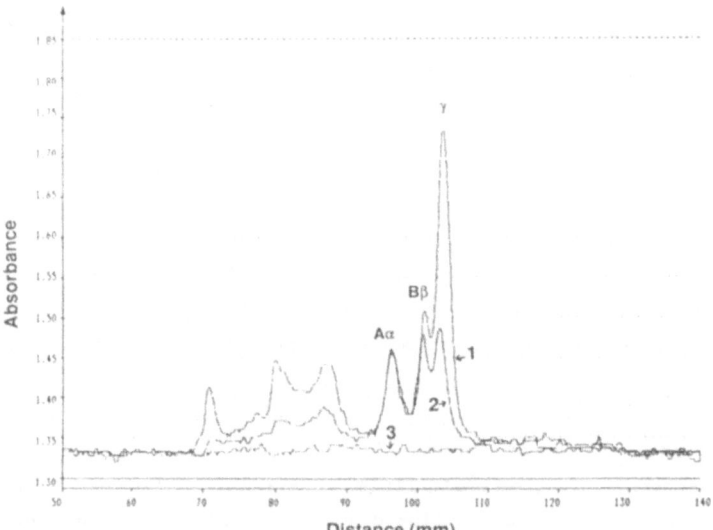

Fig. 1. Densitometric scanning of autoradiogram showing the specific binding of ^{125}I-GPRVVERHK to the γ chain of human fibrinogen. Cross-linking experiments performed in the absence (curve 1) and in the presence of various concentrations of the unlabeled peptide; a 40-fold (curve 2) and 100-fold (curve 3) molar excess was used.

the peptide, autoradiograms were subjected to densitometric scanning. At 40-fold molar ratio excess (Fig. 1: curve 2), GPRVVERHK produced 65% and 1% inhibition of the binding of ^{125}I-GPRVVERHK to the γ and Bβ chains, respectively. At the same concentration there was no inhibition observed in the case of the Aα chain. A 100-fold molar excess of nonlabeled GPRVVERHK (Fig. 1:curve 3) abolished the cross-linking of the radioactive peptide to all three polypeptide chains.

With the above data providing clear evidence for specific cross-linking of the GPRVVERHK to a relevant site in the γ chain we sought to localize this site within the plasmic cleavage products of fibrinogen. For this purpose ^{125}I-GPRVVERHK was cross-linked to fibrinogen. Then fibrinogen was digested with plasmin. Samples of plasmic digests were analyzed by SDS-PAGE under nonreducing (11% acrylamide gels) and reducing (15% acrylamide gels) conditions. Gels were stained with Coomassie blue, destained, dried, and subjected to autoradiography. Autoradiograms clearly indicate that after cross-linking of ^{125}I-GPRVVERHK to fibrinogen and digestion

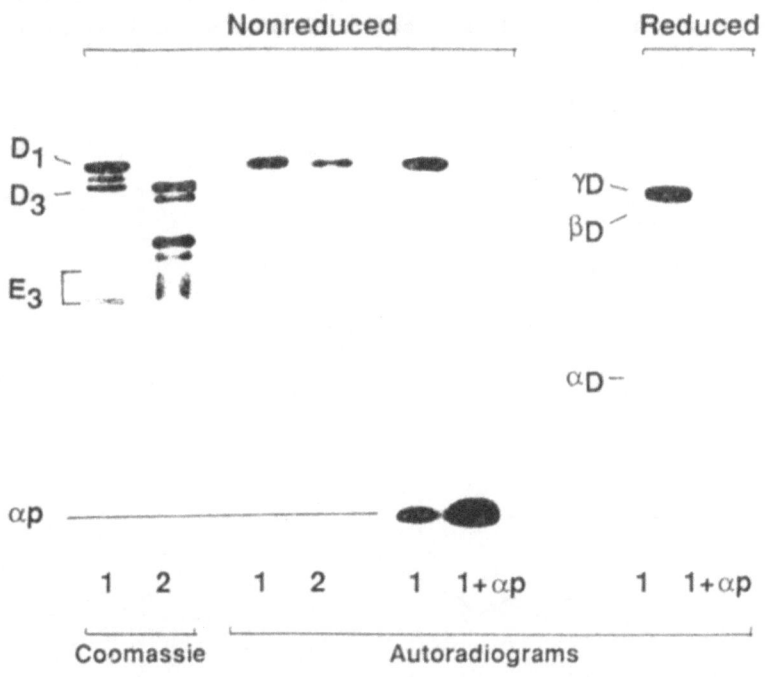

Fig. 2. The recovery of the ^{125}I-GPRVVERHK in plasmic cleavage products of fibrinogen shown by SDS PAGE. Fibrinogen cross-linked with ^{125}I-GPRVVERHK was digested at 37° C with plasmin (0.5 CTA units) in the absence of Ca^{2+} ions for two hours (lanes 1) or 24 hours (lanes 2). In some experiments, ^{125}I-GPRVVERHK was cross-linked with fibrinogen in the presence of a 40-fold molar excess of unlabeled GPRVVERHK (lanes 1+AP). Samples of digests were separated by SDS PAGE under nonreducing and reducing conditions.

with plasmin, all detectable radioactivity migrated in the position of FgD₁ and its
γ chain remnant (Figure 2).   This cross-linking was inhibited in almost 100% by a
40-fold molar excess of nonlabeled GPRVVERHK.   Further digestion of FgD₁ to FgD₂
and FgD₃ caused the release of the radioactive peptide.   Neither FgD₂ and FgD₃ or
their γ chain remnants contained cross-linked ¹²⁵I-GPRVVERHK.

To narrow down the location of cross-linking site for GPRVVERHK within the γ
chain, FgD₁ containing ¹²⁵I-GPRVVERHK was isolated from a 2 hour-plasmic digest of
fibrinogen by affinity chromatography using antibodies to Fg D₁ immobilized on Sepharose 4B.   Then, fragment D₁ was treated with EGTA (20 mM) and further digested
with plasmin (0.1 CTA unit) for 24 hours at 37°C.   The digest was separated by HPLC
on the C₁₈ column (Fig. 3).   Fractions corresponding to each peak were analyzed for
radioactivity using a γ counter.   The radioactive component was found in peak
19.07.   Amino acid sequencing of the major radioactive cleavage product released
from FgD₁ by plasmin yielded one major sequence corresponding to that of γ fragment
extending from residue 357 to 373.   The second   minor sequence obtained could be
read for 15 cycles and gave interpretable signals which could be identified with
the peptide fragment γ₃₀₂₋₃₂₁.   It also showed the presence of GPRVV sequence.

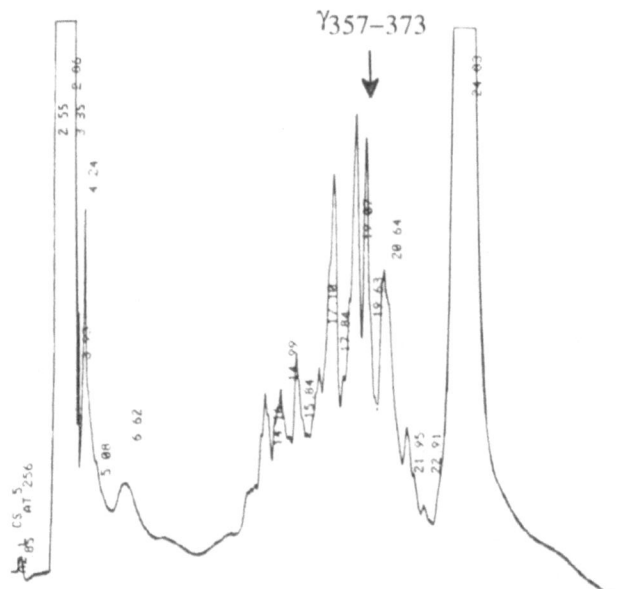

Fig. 3.   Isolation of the γ chain fragment cross-linked with ¹²⁵I-GPRVVERHK by
reverse phase HPLC using a C₁₈ column (Waters).   Plasmic digest of fragment D₁ was
directly applied to the C₁₈ column and a 0.1% trifluoroacetic acid (solvent A) and
0.1% trifluoroacetic acid in acetonitrile (solvent B) gradient was used as eluant.
A linear gradient of 0 to 50% solvent B in 40 minutes with a flow rate of 1 ml/min
was used.   The column effluent was monitored at 212 nm.   Radioactivity of isolated
fractions was detected by the γ counter and the peak 19.07 was found predominantly
to contain a radioactive component.

DISCUSSION

Recent studies focused on defining areas of the fibrin monomer molecule that participate in polymerization indicate that polymerization sites have a complex structure.[6] It means that they are not formed simply by a linear sequence of amino acid residues but they require for complete expression the presence of several not necessarily contiguous peptide segments in close proximity. Such a complex structure explains why short peptide fragments derived from polymerization sites, for example Gly-Pro-Arg-Pro, only weakly bind to the complementary polymerization sites present in the Fragment D_1.[8] Despite a weak inhibitory activity, short peptides can be very useful in detection of complementary interacting structures in intact protein macromolecules.[9] Hence, in order to identify protein sequences which provide the complementary binding site "a" in intact fibrinogen molecule, we used peptide GPRVVERHK derived from the NH_2-terminus of fibrin α chain containing the active tripeptide Gly-Pro-Arg.

A summary of the data localizing the GPRVVERHK cross-linking site in fibrinogen molecule is schematically illustrated in Figure 4. The binding of ^{125}I-GPRVVERHK to the γ chain was specifically inhibited by a 40-fold molar excess of nonlabeled peptide. The radioactive peptide remained in FgD₁. Further digestion of FgD₁ to FgD₂ associated with the release of $\gamma_{367-411}$ resulted in a loss of the cross-linked ^{125}I-GPRVVERHK. Amino acid sequencing of the isolated peak revealed one major peptide with sequence corresponding to the $\gamma_{357-373}$.

It is tempting to speculate that polymerization site "a" consists of two finger-shaped regions of the γ chain separated by disulphide bridge $\gamma_{326-\gamma339}$ (Fig. 4).

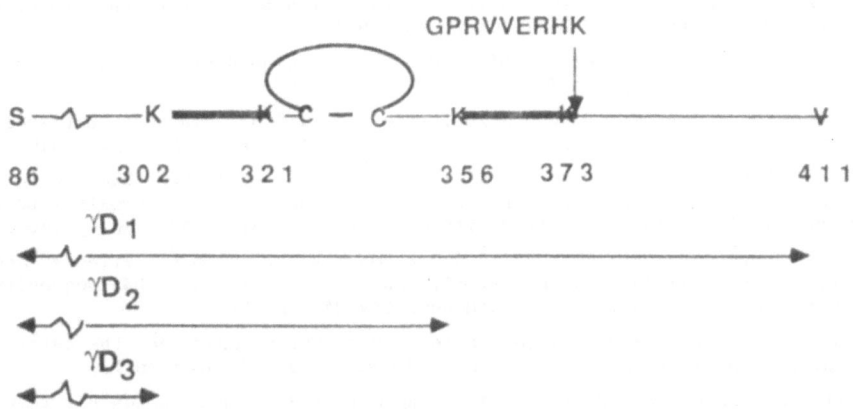

Fig. 4. Diagrammatic representation of the results establishing the location of the GPRVVERHK cross-linking site within the γ chain of fibrinogen. ^{125}I-GPRVVERHK become cross-linked to lysine at the position 373. Thus the peptide is present in FgD₁ but not in D₂ or D₃. γ_{D1}, γ_{D2}, and γ_{D3} correspond to the γ chain remnant of fragment D₁ (86-411), D₂ (86-356), and D₃ (86-302), respectively.

Such a complex structure of the polymerization site "a" is in an excellent agreement with following observations: 1) conversion of FgD_1 to FgD_2 resulting in the release of $\gamma_{357-411}$ is associated with the loss of anticlotting activity.[3], 2) Ca^{2+} ions associated with the sequence $\gamma_{303-321}$ protect the peptide bond $\gamma356/357$ against digestion by plasmin and influence the rate of fibrin monomer polymerization.[3], 3) Regions $\gamma_{308-352}$ and $\gamma_{363-385}$ are highly conserved in distant species[10], 4) The cleavage of disulphide bridge $\gamma_{326}-\gamma_{339}$ separating both γ chain segments causes a significant prolongation of thrombin clotting time[11]. It also explains a critical effect of natural mutations found in congenital polymerization-defective dysfibrinogens to occur at the positions 275, 292 , 308, 310, 329, and 330 which may directly or undirectly cause dissociation of both segments and thus change the expression of the polymerization site.[12-14]

REFERENCES

1. Bettelheim F.R. and Bailey K. (1952) The products of the action of thrombin on fibrinogen. Biochim. Biophys. Acta 9: 578-579.

2. Southan C., Thompson E., Panico M., Etienne T., Morris H.R. and Lane D.A. (1985) Characterization of peptides cleaved by plasmin from the C-terminal polymerization domain of human fibrinogen. J. Biol. Chem. 260: 13095-13101.

3. Varadi A. and Scheraga H.A. (1986) Localization of segments essential for polymerization and for calcium binding in the γ-chain of human fibrinogen. Biochemistry 25: 519-528.

4. Kuyas C., Sigrist H. and Straub P.W. (1987) Localization of fibrin polymerization sites by photoaffinity. Thromb. Haemost. 58: 287 (abstract).

5. Shimuzu A. and Doolittle R.F. (1989) Identification of fibrin polymerization site by photoaffinity labeling. Thromb. Haemost. 62: 70 (abstract).

6. Cierniewski C.S., Kloczewiak M. and Budzynski A.Z. (1986) Expression of primary polymerization sites in the D domain of human fibrinogen depends on intact conformation. J. Biol. Chem. 261: 9116-9121.

7. Laemmli U.K. (1970) Cleavage of structural proteins during the assembly of the head of bacteriophage T4. Nature 227: 680-685.

8. Laudano A.P., Cottrell B.A. and Doolittle R.F. (1983) Synthetic peptides modelled on fibrin polymerization sites. Ann. N.Y. Acad. Sci. 408: 315-329.

9. D'Souza S.E., Ginsberg M.H., Burke T.A. and Plow E.F. (1990) The ligand binding site of the platelet integrin receptor GPIIb-IIIa is proximal to the second calcium binding domain of its α subunit. J.Biol.Chem. 265: 3440-3446.

10. Strong D.D., Moore M., Cottrell B.A., Bohonus V.I., Evans B., Riley M. and Doolittle R.F. (1985) Lamprey fibrinogen γ chain: cloning, cDNA sequencing and general characterization. Biochemistry 24: 92-101.

11. Procyk R. and Blomback B. (1987) Role of disulphide bonds near the calcium binding sites in fibrinogen. Thromb. Haemost. 58: 38 (abstract).

12. Bantia S., Mane S.M., Bell W.R. and Dang C.V. (1990) Fibrinogen Baltimore I: Polymerization defect associated with a $\gamma^{292}Gly\rightarrow Val$ (GGC$\rightarrow$GTC) mutation. Blood 76: 2279-2283.

13. Bantia S., Bell W.R. and Dang D.V. (1990) Polymerization defect of fibrinogen Baltimore III due to a $\gamma Asn^{308}\rightarrow Ile$ mutation. Blood 75: 1659-1663.

14. Muramatsu S., Mimuro J., Mackawa H., Sakata Y., Matsuda M., Yoshitake S., Okuma M. (1991) Gene analysis of abnormal fibrinogens characterized by a mutation in the γ chain. Thromb. Haemost. 65: 821 (abstract).

Molecular Interaction Between Thrombin and Thrombomodulin

in The Protein C Pathway

Koji Suzuki: Department of Molecular Biology on Genetic Disease,
Mie University School of Medicine, Tsu-city, Mie 514, Japan.

INTRODUCTION

The protein C pathway is one of the most important regulatory systems of the intravascular blood coagulation, because the patients with hereditary or acquired deficiency of protein C or protein S, both are major components in this pathway, often result in very severe recurrent venous thrombosis (1-4). Protein C pathway, which is composed of several plasma and cellular components (5), consists of three important processes. The first step is the activation of protein C by thrombin which is bound to thrombomodulin (TM), a receptor for thrombin on the surface of endothelial cells. The second step is the activated protein C (APC)-catalyzed inactivation of coagulation cofactor proteins, Factors Va and VIIIa. This inactivation by APC is accelerated by protein S, which is another vitamin K-dependent plasma protein. The third step is the neutralization of APC by plasma serine protease inhibitors: protein C inhibitor and α_1-antitrypsin.

In the activation of protein C, TM markedly accelerates the thrombin-catalyzed activation of protein C, and moreover it inhibits the procoagulant activities of thrombin, such as fibrinogen clotting (6), activation of Factor V (6), Factor XIII, and platelets (7), and proteolytic inactivation of protein S. In this event, there are at least two kinds of molecular assemblies: One is the interaction between thrombin and TM, and the other is the interaction between protein C and TM. The present paper describes our studies on the molecular interaction between thrombin and TM in the process of protein C activation.

I. STRUCTURE OF TM

TM is a single-chain glycoprotein with M_r approximately 78,000 (8). The primary structure of human TM was revealed by analyses of its cDNA clones (9). The TM precursor consists of a mature protein region of 557 amino acid residues and a signal peptide region of 18 residues. Mature TM can be tentatively divided into five domains as shown in Fig. 1 (9). The first domain is the extracellular NH_2-terminal region, which contains a region homologous to some kinds of animal lectins. The second domain contains 36 Cys residues, and it is composed of 6 consecutive EGF-like structures. The third domain is rich in Ser, Thr and Pro residues, and probably rich in sites for O-glycosylation including glycosaminoglycans. The fourth domain is a hydrophobic putative transmembrane domain. The fifth domain, situated at the COOH-terminal end of the molecule, is located on the cytoplasmic side of the plasma membrane. A Ser residue in this domain of mouse hemangioma TM was found to be phosphorylated in response to

Key words: Thrombomodulin, Thrombin, Protein C pathway, Molecular interaction, Epidermal growth factor domain.

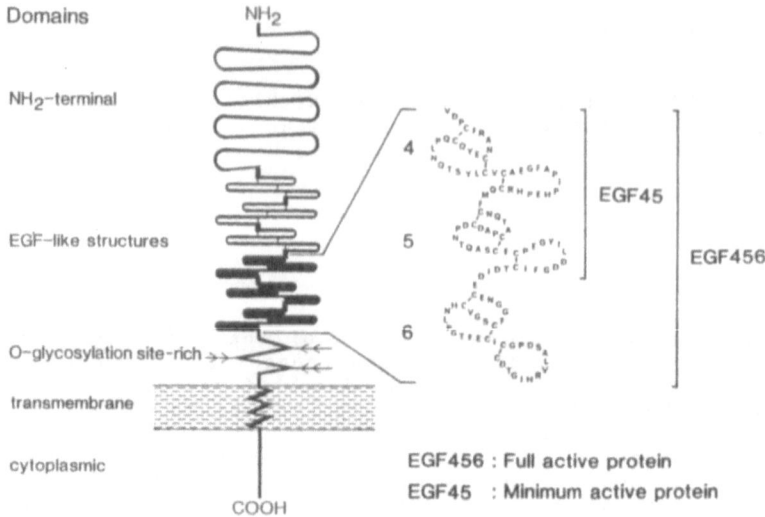

Fig. 1. Schematic model of human TM and amino acid sequence
corresponding to the 4th to 6th EGF domains which
serves the binding sites for thrombin and protein C.

treatment with a protein kinase C stimulant (10).

The gene for human TM, located at chromosome 20p12-cen and 3.7
kilobases in length, is free of introns in both the coding and the
noncoding region (11). This is an unusual situation among eukaryotic
genes, since only a few without introns have been found, such as the
genes for eukaryotic mitochondrial proteins, human α- and β-inter-
feron and angiogenine.

II. MOLECULAR INTERACTION BETWEEN THROMBIN AND TM

1. Interaction Site for Thrombin in TM

Esmon and coworkers reported that an elastase-digested fragment
from rabbit TM, which contains six EGF domains, had cofactor activity
for the thrombin-catalyzed activation of protein C (12). Similar
results were obtained by using recombinant human TM proteins composed
of six EGF domains (13). The essential region necessary for full
cofactor activity was found in a recombinant protein composed of three
consecutive EGF domains, the 4th, 5th and 6th domains (14). A cyano-
gen bromide fragment-3 from rabbit TM composed of the 5th and 6th EGF
domains had the ability to bind to thrombin, but could not activate
protein C (15).

We investigated more precise interaction site for thrombin within
TM by using several recombinant mutant proteins which were missing one
or more EGF domains in the native TM using human TM cDNA as illustrated in Fig. 2 (16). D123 protein is composed of three extracellular
domains. EGF1-6 protein contains six EGF domains. EGF456 protein
contains the 4th, 5th and 6th EGF domains. EGF45 and EGF56 proteins
are composed of the 4th and 5th EGF domains, and the 5th and 6th EGF

domains, respectively.
     First, the effects of mutant proteins on the thrombin-catalyzed activation of protein C in the presence of 2 mM Ca^{2+} ions were investigated. As shown in Fig. 3, EGF1-6 and EGF456 proteins had full cofactor activity as well as D123 and intact TM. Moreover, EGF45 protein possessed about one-tenth of the activity of E456. However, EGF56 protein did not show any cofactor activity. From these results, EGF45 consisting of the 4th and 5th EGF domains was suggested to be minimumly active for the expression of the cofactor activity.

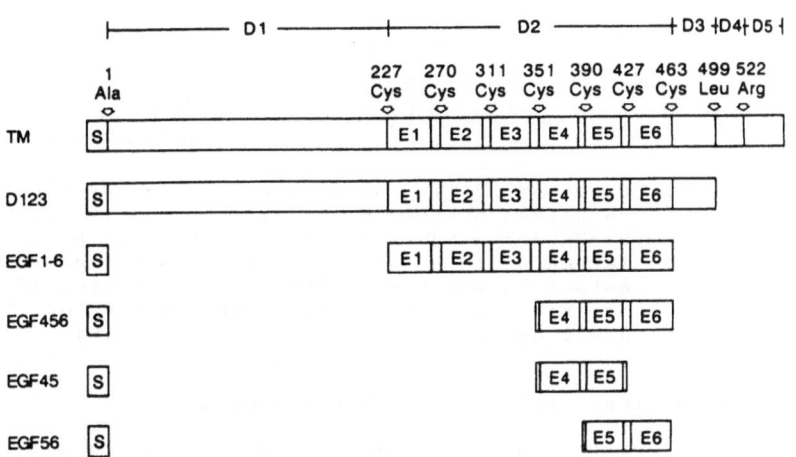

Fig. 2. Schematic representation of deletion mutants of human TM.
TM indicates the precursor of the entire protein. D123
encodes a TM protein, terminated at residue 498, which
is devoid of domains 4 and 5. The other four deletion
mutants were constructed on D2 involving six EGF domains.
S, signal peptide; D1, domain 1 (NH$_2$-terminal domain);
D2, domain 2 (a domain containing six EGF domains); D3,
domain 3 (O-glycosylation site-rich domain); D4, domain
4 (transmembrane domain); D5, domain 5 (cytoplasmic
domain). E1 to E6 denote the first to sixth EGF domains,
respectively.

Then, we studied the effects of both EGF45 and EGF56 proteins on binding of thrombin to TM fixed in microwell plates to elucidate which of the EGF domains serves the binding site for thrombin. Both EGF45 and EGF56 proteins competitively inhibited the binding of thrombin to TM, suggesting that thrombin binds to the 5th EGF domain of TM.
     Furthermore, the peptides corresponding to the sequence residues of the first, second and third loop structures of the 5th EGF domains were synthesized, and the effects of these peptides on thrombin binding to TM were examined. Then it was found that a peptide with a sequence corresponding to the third loop structure specifically inhibited the thrombin binding to TM. These findings indicate that the latter half structure of the 5th EGF domain plays a role as the binding site for thrombin.

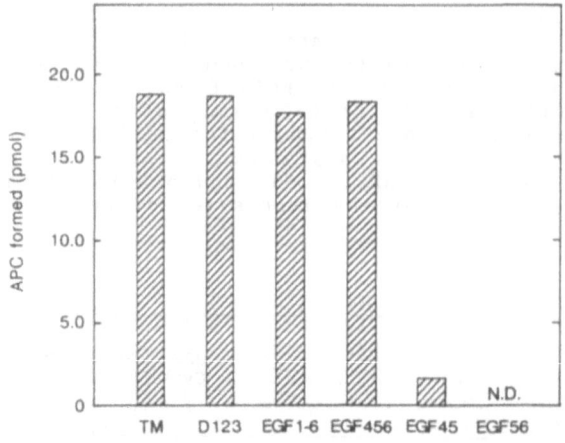

Fig. 3. Cofactor activty of mutant TM proteins for thrombin-
catalyzed protein C activation. To 100 μl of TBS with
2 mM $CaCl_2$ and 0.1% BSA were added 10 μl of 15 nM
thrombin, 10 μl of 3.2 μM protein C, and 10 μl of 10 nM
mutant TM protein. After 1 h incubation at 37 °C, APC
activity was determined using BOC-Leu-Ser-Thr-Arg-MCA.
N.D., nondetectable activity. (Ref. 16).

2. Interaction Site for TM within Thrombin

TM stimulates thrombin-catalyzed activation of protein C, and it
also inhibits the procoagulant activities of thrombin itself such as
fibrinogen clotting, Factor V activation, and platelet activation.
This TM-induced conversion of substrate specificity of thrombin was in
part demonstrated by Musci et al. (17), in which they show that it
results from a conformational change within or around the active
center of thrombin in the presence of TM by using fluorosulfonyl
spin-label inhibitors.
α-thrombin is autocatalytically or tryptic-catalytically converted into β-thrombin and γ-thrombin. Although the amidolytic activity
of β- and γ-thrombin is almost completely conserved, both catalytic
products show very low abilities for TM-dependent protein C activation, and also have very low affinities for TM, in comparison with
those achieved by α-thrombin. This may suggest that the interaction
site for TM is involved in two peptide regions of the B-chain, which
are cleaved during the conversion of α-thrombin into β- and γ-thrombin. A region comprised of residues Arg-62 to Arg-73 of the B-chain
was suggested to interact with TM as well as with fibrinogen or hirudin, based on the result that a monospecific antibody which was raised
from a peptide composed of residues Arg-62 to Arg-73 inhibited the
thrombin-TM-induced protein C activation (18).
We investigated the interaction site for TM within thrombin by
analyzing an epitope for a monoclonal antibody for thrombin, called
MT-6, which inhibited TM-dependent protein C activation by thrombin
and also thrombin binding to TM (19). DIP-α-thrombin was digested
with Staphylococcus aureus V8 protease, and degradation products were
detected on nitrocellulose membrane by both ligand-blotting using TM
and the immunoblotting using the monoclonal antibody for thrombin.

A peptide fragment, which bound to both TM and the antibody, was isolated from reduced and S-pyridylethylated DIP-α-thrombin by V8 protease digestion followed by reverse-phase HPLC. A fraction positive for both TM and the antibody contained an 8.5-KDa fragment. This fragment was composed of three peptides which were linked by two disulfide bonds, then it was further reduced and S-pyridylethylated, and subjected to HPLC. The first peak positive for both TM and the antibody was isolated, and amino acid sequence of this peptide was determined. The sequence of this peptide coincided to the sequence residues Thr-147 to Asp-175 of the B-chain of thrombin (Fig. 4). Then a peptide with a sequence corresponding to the residues Thr-147 to Ser-158, which was presumed to be located on the surface of thrombin molecule, was synthesized and it was determined whether this peptide inhibits thrombin binding to TM. The peptide directly inhibited the thrombin binding to TM fixed in the microwell plate (Fig. 4). These findings suggest that the interaction site for TM within thrombin is located in the sequence residues Thr-147 to Ser-158 of the B-chain.

Isolated peptide   [147]TWTANVGKGQPSVLQVVNLPIVERPVCKD[175]

Synthetic peptide   [147]TWTANVGKGQPS[158]

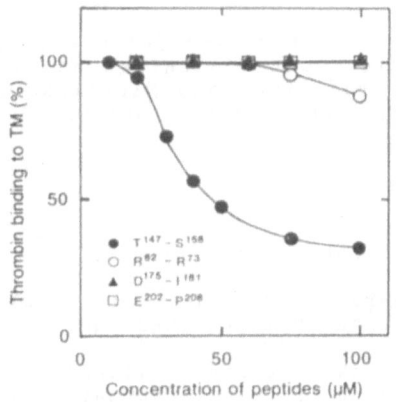

Fig. 4. Amino acid sequence of the peptide isolated as the epitope for a monoclonal anti-thrombin-IgG, MT-6, and that of the synthetic peptide. Figure shows the effects of peptides corresponding to the sequence residues Thr-147 to Ser-158 and other regions in the B-chain of thrombin on the binding of thrombin to TM fixed in microwell plate.

We also found that this peptide specifically and directly inhibited procoagulant activities of thrombin, such as fibrinogen clotting, Factor V activation and platelet activation (20). Moreover, this peptide blocked thrombin inhibition by hirudin, but did not thrombin inhibition by antithrombin III (20). These findings suggest that the binding site for TM in thrombin is shared for the site for fibrinogen, Factor V, platelet or hirudin.

The region of the binding site for TM is presumed to be located at the edge of the active center pocket of thrombin by the analysis of the three-dimensional structure of thrombin which was predicted from the structure of trypsin by computer graphics. The binding of TM to this region probably disturb the interaction between thrombin and fibrinogen, Factor V and/or platelets by steric hindrance. In addition to this, the binding of TM may also lead to conformational changes within the active center of thrombin to facilitate the protein C activation.

III. CONCLUSION

On the activation of protein C by thrombin and TM, a region in the latter half of the 5th EGF domain of TM interacts with a region corresponding to the sequence rsidues Thr-147 to Ser-158 located near the active center of thrombin, and TM may induce the conformational change in thrombin to enhance the velocity of the thrombin-catalyzed protein C activation. At the same time TM blocks the interaction between thrombin and procoagulant substrates such as fibrinogen, Factor V and platelets. Although study on the interaction between protein C and TM is not described here, we recently found that a binding site for protein C is located at a region in the 4th EGF domain of TM. Thus, EGF domains in TM play an important role in the activation of thrombin-catalyzed protein C activation.

Ackowledgements: I deeply thank the following collaborators performing the present studies; Junji Nishioka and Tatsuya Hayashi in Mie University School of Medicine, and Michitaka Zushi, Komakazu Gomi, Gouichi Honda and Shuji Yamamoto in Asahi Chemical Industry.

REFERENCES

1. Esmon CT (1987) Science 235: 1348-1352

2. Broekmans AW, Conard F (1988) In: Bertina RM (ed) Protein C and Related Proteins. Churchill Livingstone, Edinburgh, pp. 160-181.

3. Briet E, Broekmans AW, Engesser L (1988) In: Bertina RM (ed) Protein C and Related Proteins. Churchill Livingstone, Edinburgh, pp. 203-212.

4. Pabinger I, Lechner K (1988) In: Bertina RM (ed) Protein C and Related Proteins. Churchill Livingstone, Edinburgh, pp. 213-225.

5. Stenflo J (1988) In: Bertina RM (ed) Protein C and Related Proteins. Churchill Livingstone, Edinburgh, pp. 21-54.

6. Esmon CT, Esmon NL, Harris KW (1982) J Biol Chem 257: 7944-7947

7. Esmon NL, Carroll RC, Esmon CT (1983) J Biol Chem 258: 12238-12242

8. Esmon NL, Owen WG, Esmon CT (1982) J Biol Chem 257: 859-864

9. Suzuki k, Kusumoto H, Deyashiki Y, Nishioka J, Maruyama I, Zushi M, Kawahara S, Honda G, Yamamoto S, Horiguchi S (1987) EMBO J 6: 1891-1897

10. Dittman WA, Kumada T, Sadler JE, Majerus PW (1988) J Biol Chem 263: 15815-15822

11. Shirai T, Shiojiri S, Ito H, Yamamoto S, Kusumoto H, Deyashiki Y, Maruyama I, Suzuki K (1988) J Biochem 103: 281-185

12. Kurosawa S, Galvin JB, Esmon NL, Esmon CT (1987) J Biol Chem 262: 2206-2212

13. Suzuki K, Hayashi T, Nishioka J, Zushi M, Honda G, Yamamoto S (1989) J Biol Chem 264: 4872-4876

14. Zushi M, Gomi K, Yamamoto S, Maruyama I, Hayashi T, Suzuki K (1989) J Biol Chem 264: 10351-10353

15. Kurosawa S, Stearns DJ, Jackson KW, Esmon CT (1988) J Biol Chem 263: 5993-5996

16. Hayashi T, Zushi M, Yamamoto S, Suzuki K (1990) J Biol Chem 265: 20156-20159

17. Musci G, Berliner LJ, Esmon CT (1988) Biochemistry 27: 769-773

18. Noe G, Hofsteenge J, Rovelli G, Stones SR (1988) J Biol Chem 263: 11729-11735

19. Suzuki K, Nishioka J, Hayashi T (1990) J Biol Chem 265: 13263-13267

20. Suzuki K, Nishioka J (1991) J Biol Chem (in press)

HEPARIN-BINDING PROPERTY OF HUMAN PROTEIN C

TAKEHIKO KOIDE[1] AND YOSHIAKI KAZAMA[2]
[1]Department of Life Science, Faculty of Science, Himeji Institute of Technology, Kamigori-cho, Hyogo, 678-12, Japan.
[2]Department of Pathology, University of New Mexico School of Medicine, Albuquerque, New Mexico 87131, USA.

INTRODUCTION

Heparin is well known to interact with various proteins in human plasma [1] which include most of the regulators of coagulation and fibrinolysis such as antithrombin III, heparin cofactor II, protein C inhibitor (PCI), histidine-rich glycoprotein, vitronectin and thrombospondin, and coagulation factors such as factor IX, factor XI, factor XII, prekallikrein, kininogen, fibrinogen, von Willebrand factor and fibronectin.

Protein C is the zymogen form of a vitamin K-dependent serine protease present in plasma [2]. Protein C is composed of two chains: the light-chain consists of γ-carboxyglutamic acid (Gla) domain and two epidermal growth factor-like (EGF) domains, and the heavy-chain consists of activation peptide and catalytic domain [3]. It is activated by a thrombin-thrombomodulin complex on the endothelial cell surface and activated protein C (APC), in conjunction with protein S, functions as a regulator of coagulation and also as a stimulator of fibrinolysis [4].

In course of the study of neutralization by histidine-rich glycoprotein of the heparin-dependent activity of PCI to inhibit APC, we found the strong negative effects of Ca^{2+} on the heparin neutralization by histidine-rich glycoprotein, and it was suggested that this phenomenon is due to the high affinity of APC for heparin [5]. To our knowledge, however, the interaction between heparin and protein C or APC has not been studied at all. Therefore, in the present study, we examined the heparin-binding properties of protein C and APC.

In this paper, we demonstrate that protein C, and more strongly APC, interacts with heparin, which is enhanced in the presence of Ca^{2+}, and also propose a specific region of protein C as a heparin-binding site.

MATERIALS AND METHODS

Preparation of Gla-Domainless Protein C and Gla-Domainless APC

Human protein C and APC were kindly provided by Dr Walter Kisiel (University of New Mexico). Gla-domainless protein C (GD-PC) was prepared by a limited proteolysis with Sepharose-bound chymotrypsin, essentially according to the method of Esmon et al. [6, and Gla-domainless APC (GD-APC) was prepared by activating GD-PC with a thrombin-thrombomodulin complex.

Affinity Chromatography on Heparin-Sepharose

Affinity chromatography on heparin-Sepharose was performed at 25 °C using an HR 5/10 (0.5 x 10 cm) Pharmacia-LKB column mounted on a Pharmacia fast protein liquid chromatography (FPLC) system. The column was equilibrated with 50 mM Tris-HCl and 0.1 M NaCl, pH7.4, containing 20 μM EDTA or 1 mM $CaCl_2$.

Amino Acid Sequence Determination.

The amino acid sequence was determined on a Shimadzu PSQ-1 System, a fully automated gas-phase protein sequencer. The PTH derivatives were identified by on-line PTH-analyzing system, with an isocratic separation using a Wakopak WS-PTH column (4.6 x 250 mm) [7].

KEYWORDS: Protein C; Activated protein C; Ca^{2+}; Heparin-binding fragment; Heparin-binding site.

RESULTS

Comparison of the Heparin-Binding Abilities of Protein C, APC, GD-PC and GD-APC in the Presence or Absence of Ca²⁺.

Figure 1 shows the heparin-binding properties of protein C (shown as PC), APC and their Gla-domainless derivatives, GD-PC and GD-APC. Both protein C and APC bind to a heparin-Sepharose, and eluted from the column in 26.7 min and in 31.5 min, respectively, after a gradient elution (0.1 - 1.0 M NaCl) was started, indicating that the affinity of protein C for heparin was increased by an activation to APC. The elution time of protein C and APC in the presence of EDTA was essentially the same as that in the absence of EDTA, indicating that metal ions are not required for the binding. In the presence of 1 mM Ca²⁺, however, the affinity of APC for heparin was significantly enhanced, delaying the elution time from 31.5 min to 38.3 min, while that of PC was unchanged. This Ca²⁺-dependent enhancement of heparin-binding ability of APC suggests that the Gla domain in APC which is one of the Ca²⁺-binding sites in protein C, may be involved in the interaction with heparin.

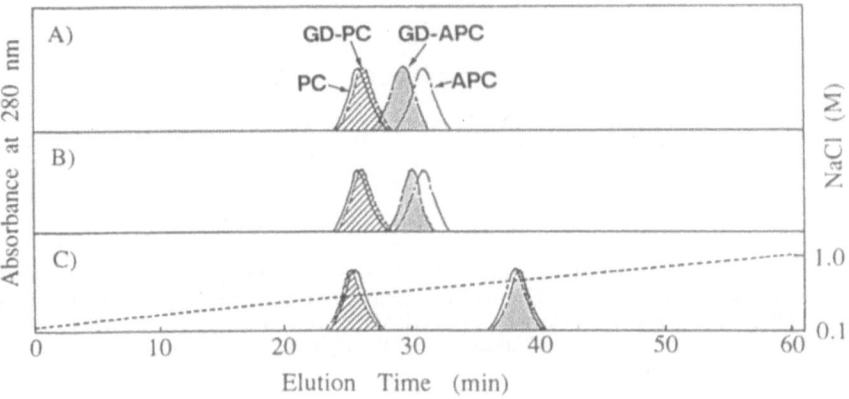

Fig. 1. Comparison of the heparin-binding abilities of protein C, APC, and their Gla-domainless derivatives. Affinity chromatography of protein C, APC, GD-PC and GD-APC on heparin-Sepharose was performed at 25 °C using a column (0.5 x 10 cm) mounted on an FPLC system. The column was equilibrated with, A) 50 mM Tris-0.1 M NaCl, pH7.4, B) 50 mM Tris-0.1 M NaCl, pH7.4, containing 20 µM EDTA, or C) 50 mM Tris-0.1 M NaCl, pH7.4, containing 1 mM CaCl₂. Each protein was applied separately and eluted with a linear gradient of NaCl concentration (0.1 - 1.0 M in 60 min) at a flow rate of 0.2 ml/min. The elution time indicates the time after a gradient elution was started. The height and shape of each protein peak is arbitrary.

In order to examine the contribution of Gla domain to the heparin-binding properties of protein C and APC, the heparin-binding abilities of GD-PC and GD-APC were compared with those of protein C and APC, respectively. As shown in Fig. 1, it is evident that both GD-PC (hatched peak) and GD-APC (shaded peak) retain essentially the same affinity as protein C and APC, respectively, for heparin. Furthermore, GD-APC showed the same Ca²⁺-dependent enhancement as APC in its heparin binding. From these results, Gla domain in protein C or APC is not required for the Ca²⁺-dependent enhancement of heparin binding ability.

Preparation of the Heparin-Binding Fragment by Limited Proteolysis

In order to investigate the heparin-binding site in protein C, the protein was digested with Sepharose-bound trypsin at 4 °C for 5 - 60 min. Figure 2 shows the time-dependent degradation of protein C, giving 34K, 28K and 17K fragments in 60 min. A 60-min digest was then incubated overnight at 4 °C with heparin-Sepharose, equilibrated with 50 mM Tris-HCl, pH 7.4, and 0.1 M NaCl, in the presence or absence of 1 mM Ca^{2+}. Heparin-Sepharose was then packed into an HR 5/10 column (0.5 x 10 cm) on an FPLC system, and after washing the column with an equilibration buffer, the bound fragment(s) was eluted with a gradient of 0.1 M to 1.0 M NaCl in 60 min. The heparin-bound fragment(s) was eluted at a concentration of about 0.55 M NaCl (elution time of 30.0 min) in the absence of Ca^{2+}, and at a concentration of about 0.75M NaCl (elution time of 39.1 min) in the presence of 1 mM Ca^{2+}, both of which are quite similar to those of APC and GD-APC, indicating that the heparin-binding property of APC is fully retained in this trypsin digest.

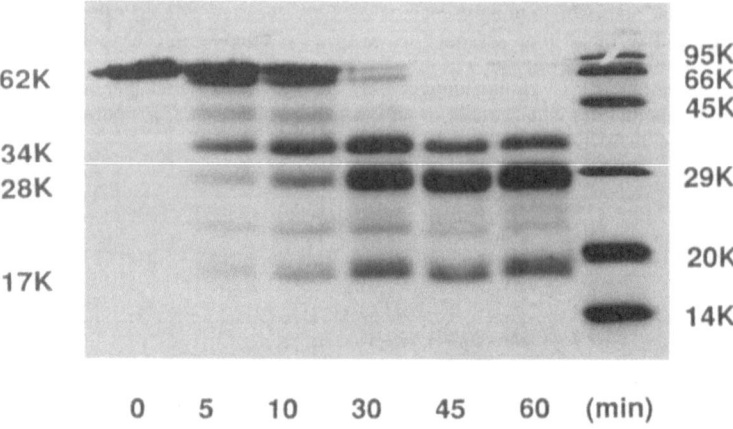

Fig. 2. Degradation of protein C by a digestion with Sepharose-bound trypsin. Protein C was digested at 4 °C with Sepharose-bound trypsin. Aliquots of samples were withdrawn at different intervals and subjected to SDS-polyacrylamide gel electrophoresis (PAGE) (10%) without any reduction. A sample on the left hand side lane is intact protein C and those on the right hand side lane are standard molecular markers.

Figure 3 shows the SDS-PAGE patterns of heparin-bound/unbound trypsin digests of protein C. A 34K band was identified as a heparin-binding fragment (34K fragment), and 28K and 17K bands as heparin-nonbinding fragments (28K and 17K fragments, respectively).

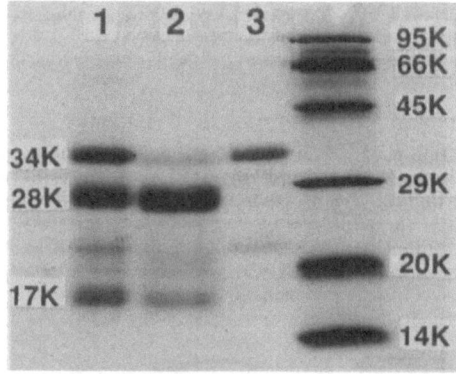

Fig. 3. SDS-PAGE of the heparin-bound/unbound trypsin digests of protein C. A 60-min trypsin digest of protein C: before application to heparin-Sepharose (lane 1), not bound to heparin-Sepharose (lane 2), and bound to heparin-Sepharose (lane 3). Samples in the right hand side lane are standard protein markers.

Isolation and Characterization of the Heparin-Binding Fragment of Protein C.

To further characterize the heparin-binding 34K fragment, it was reduced, carboxymethylated, and gel-filtered on Superose 12 prepacked in a Pharmacia HR 10/30 column (1.0 x 30 cm), equilibrated with 9 % formic acid. The amino-terminal amino acid sequence analysis of one isolated 20 kDa peptide gave a sequence of Leu-Ile-Asp-Gly-Lys-Met-Thr-Arg-Arg-Gly-, which corresponds to the sequence of Leu13 to Gly22 in the heavy-chain of protein C (Leu1 to Gly10 in APC), indicating that Sepharose-bound trypsin hydrolyzed an Arg12-Leu bond, the same peptide bond as a thrombin-thrombomodulin complex hydrolyzes to activate protein C and release the activation peptide. From this result, together with that of amino acid analysis, it was estimated that the 20 kDa peptide consists of Leu1 through Arg145 and/or Lys153 of the heavy-chain of APC. The amino-terminal sequence analysis of another isolated peptide gave a sequence of His-Val-Asp-Gly-Asp-Gln-Cys(Cm)-Leu-Val-Leu-, which corresponds to the sequence of His44 to Leu53 in the light-chain of protein C. From this result, together with that of amino acid analysis, this peptide was estimated to consist of His44 through Lys146 of the light-chain of protein C. Thus, the 34K fragment was identified to consist of two peptide chains, one involving two EGF domains (His44 to Lys146) and the other involving the amino-terminal portion of the heavy-chain of APC (Leu1 to Arg145 and/or Lys153).

After blotting protein bands on SDS-gel to PVDF (polyvinylidene difluoride) membrane, the amino-terminal sequence of each band was determined to identify other tryptic fragments.

The 28K fragment gave one major and two minor sequences. The major sequence was determined as His-Val-Asp-Gly-Asp-Gln- ? -Leu-Val-Leu, which corresponds to His44 through Leu53 of the light-chain of protein C, and two minor sequences were estimated as Trp-Glu-Lys-Trp-Glu-Leu-Asp-Leu-Asp-Ile and Trp-Glu-Leu-Asp-Leu-Asp-Ile-Lys-Glu-Val, corresponding to those sequences starting from Trp74 and Trp77, respectively, of protein C (Trp62 and Trp65, respectively, in APC).

The 17K fragment gave one major and one minor sequences of Thr-Phe-Val-Leu-Asn-Phe- and Ile-Pro-Val-Val-Pro-His- with a molar ratio of 3 to 1. The major sequence corresponds to that starts from Thr158 and the minor one from Ile166 in the heavy-chain of protein C (Thr 146 and Ile154, respectively, in APC).

Even after a 60-min incubation, Sepharose-bound trypsin did not hydrolyze a Lys138-Phe bond in the light-chain of protein C to isolate EGF domains from the remaining, which is probably due to the steric hindrance. Therefore, we employed free TPCK-trypsin, and a 60-min digest was similarly treated with heparin-Sepharose. However, none has bound to heparin-Sepharose. Then, the TPCK-trypsin digests of protein C were examined on SDS-PAGE as shown in Fig. 4. There is no 34K band in a 60-min digest any more, but newly a 24K band can be observed.

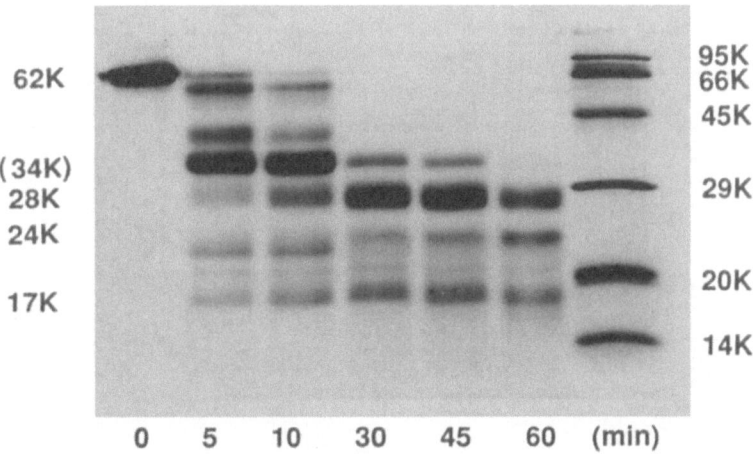

Fig. 4. Degradation of protein C by a digestion with TPCK-trypsin. Protein C was digested with TPCK-trypsin at 4 ºC. Aliquots of samples were withdrawn at different intervals and subjected to SDS-PAGE (10%) without any reduction. A sample on the left hand side lane is intact protein C and those on the right hand side lane are standard molecular markers.

After blotting protein bands on SDS-gel to PVDF membrane, the amino-terminal sequence of each band was determined to identify the tryptic fragments. The 28K and 17K fragments gave similar results described above. The amino-terminal sequence analysis of the 24K fragment gave a single sequence of His-Val-Asp-Gly-Asp-Gln- ? - Leu-Val-Leu-Pro-Leu, which corresponds to His44 through Leu55 of the light-chain of protein C. Although we could not obtain a reliable data on amino acid composition of this peptide, it is highly likely that this peptide has been produced by the tryptic cleavage of a Lys138-Phe bond in addition to the cleave of a Lys43-His bond.

Figure 5 summarizes the sites in protein C specifically hydrolyzed by Sepharose-bound trypsin and by TPCK-trypsin. Sepharose-bound trypsin was shown to hydrolyze at least four specific sites in protein C: a Lys43-His bond in the light-chain and an Arg12-Leu bond in the heavy-chain of protein C from amino-terminal sequence analyses, and a Lys146-Arg bond of the light-chain and an Arg145-Thr and/or a Lys153-Ile bond(s) of the heavy-chain of APC from amino acid analyses of the peptides isolated after reduction of the 34K fragment. Besides this, TPCK-trypsin hydrolyzed at least one additional site, i.e. a Lys138-Phe bond, separating EGF domain from the remaining as shown by Öhlin and Stenflo [8].

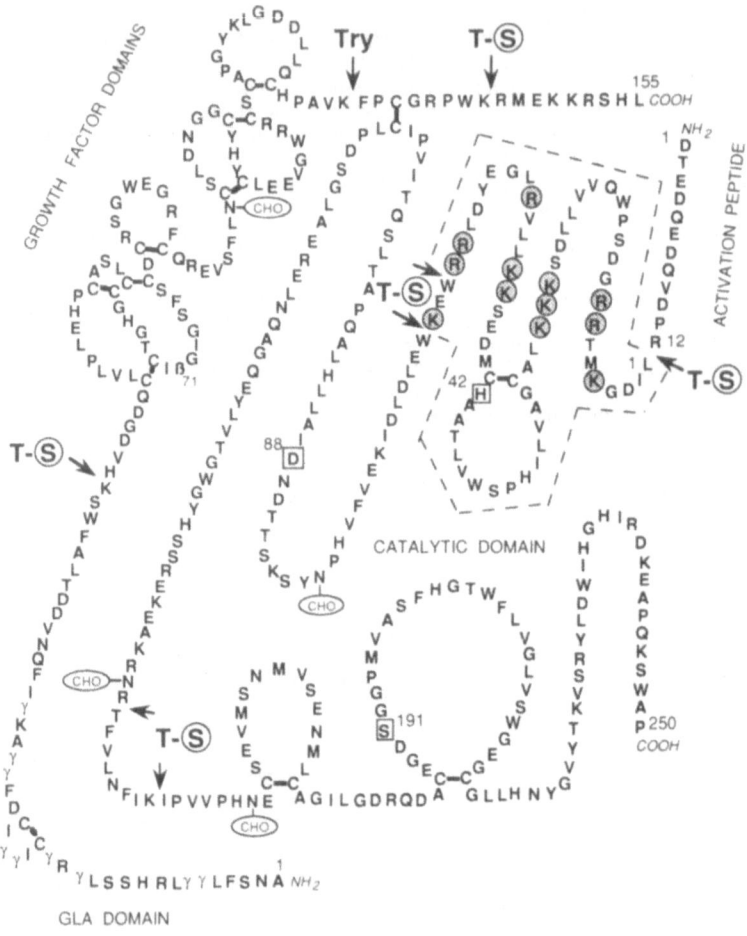

Fig. 5. Cleavage sites in protein C by Sepharose-bound trypsin and by TPCK-trypsin. The amino acid residue number of the heavy-chain is expressed as that of APC. Arrows indicate the sites identified or estimated to have been cleaved specifically by trypsin. T-Ⓢ, Sepharose-bound trypsin; Try, TPCK-trypsin. The proposed heparin-binding region is surrounded by broken lines, and basic amino acid residues in this region are shown in shaded circles.

DISCUSSION

In the present study, we have demonstrated for the first time that protein C, and more prominently its activated form, APC has an affinity for heparin. Calcium ion is not required for the binding of protein C or APC to heparin, but it enhances the affinity of APC for heparin. To our surprise, this property is independent of both Gla domain and EGF domain, both of which are known to be involved in the Ca^{2+}-binding of protein C [4].

It should be pointed out that a significant heparin-binding property of protein C is generated only after activation to APC. This is similar to thrombin whose high-affinity for heparin is not marked in prothrombin. Since the activation peptide of protein C which contains six acidic residues out of 12 amino acid residues, is exceptionally rich in negative charge, when compared with those of other vitamin K-dependent coagulation factors such as factor IX (9 acidic residues out of 34) and factor X (10 acidic residues out of 52), it may directly interrupt the interaction between heparin and the heparin-binding site of protein C in the adjacent region, or it may form an intramolecular ionic interaction with the heparin-binding site rich in positive charge (Fig. 5), as recently proposed for heparin-cofactor II [9].

The enhancement of affinity of protein C for heparin by the activation to APC may be physiologically important. First, if protein C has a high affinity for heparin, the activation of protein C by a thrombin-thrombomodulin complex may be interrupted by forming a complex of protein C and heparin or a ternary complex of protein C, heparin and PCI which also has an affinity for heparin. Secondly, it is of great advantage for a PCI-heparin complex to attack and form a complex with APC that has gotten a high affinity for heparin by activation.

From a remarkable difference in the heparin-binding ability between the 34K fragment and the 28K fragment obtained from a tryptic digest of protein C, we propose a region of Leu1 to Lys64 in the heavy-chain of APC as a heparin-binding site in protein C (Fig. 5). Protein C contains as many as 11 basic amino acid residues in this region, while other vitamin K-dependent coagulation factors such as thrombin, factor VIIa, factor IXa and factor Xa contain only four to five basic amino acid residues in the corresponding regions. In the present study, however, it was unsuccessful to obtain a fragment corresponding to this region, which probably degrades very easily after the cleavage of Arg61-Trp bond and/or Lys64-Trp bond by trypsin.

No information is available from references to explain the Ca^{2+}-dependent enhancement of heparin binding of APC. In factor IX, however, it has recently been reported that the catalytic domain contains a high Ca^{2+}-binding region, and Glu235 and Glu245 in factor IX have been proposed to form a Ca^{2+}-binding site [10]. Protein C has Glu225 (56 in the heavy-chain of APC) and Glu235 (66 in APC) at the corresponding positions to those of factor IX. Therefore, it is possible to assume that APC also has a Ca^{2+}-binding site in this region and Ca^{2+}-binding to this site induces a conformational change in APC, in GD-APC and in the 34K tryptic fragment of protein C which is more favorable for the heparin-binding.

ACKNOWLEDGEMENTS

We thank Dr. Walter Kisiel for kindly providing human PC and APC. This work was supported in part by a Grant-in-Aid for Scientific Research (01480298) and a Grant-in-Aid for Scientific Research on Priority Areas (63616512) from the Ministry of Education, Science, and Culture of Japan.

REFERENCES

1. Ofosu FA, Danishefsky I, Hirsh J (eds) Heparin and Related Polysaccharides. Structure and Activities. (1989) Ann NY Acad Sci 556.
2. Kisiel W (1979) J Clin Invest 64: 761-769
3. Foster DC, Davie EW (1984) Proc Natl Acad Sci USA 81: 4766-4770
4. Esmon CT (1989) J Biol Chem 264: 4743-4746
5. Kazama Y, Koide T (1991) Thromb Haemostas (in press)
6. Esmon NL, DeBault LE, Esmon CT (1983) J Biol Chem 258: 5548-5553
7. Tsunasawa S, Kondo J, Sakiyama F (1985) J Biochem (Tokyo) 97: 701
8. Öhlin A-K, Stenflo J (1987) J Biol Chem 262: 13798-13804
9. Ragg H, Ulshöfer T, Gerewitz J (1990) J Biol Chem 265: 5211-5218
10. Bajaj SP, Sabharwal AK, Gorka J, Birktoft JJ (1991) Thromb Haemostas 65: 749 (abstr)

A NEW THROMBOSIS MODEL WITH PHOTOCHEMICAL REACTION IN THE RAT FEMORAL ARTERY

Y. Takiguchi, H. Matsuno, T. Uematsu, K. Wada and M. Nakashima
Department of Pharmacology, Hamamatsu University School of Medicine,
Hamamatsu 431-31, Japan.

INTRODUCTION

Changes in the endothelial cells are important events in the process of intravascular thrombosis. The feasibility of studying microvascular injury and its subsequent platelet activation has been demonstrated in various animal models of thrombosis. Focal thrombosis of the cerebral vessels of rat has been achieved by irradiating the skull with green light after the systemic injection of rose bengal [1].

In the present study we applied the photochemical technique to the rat femoral artery in an attempt to clarify the mechanism of thrombus formation in the large vessels. Our photochemically induced thrombosis in the rat has similar characteristics to that in human subjects, and therefore this may be useful for better understanding of the formation of intravascular thrombus and for screening and evaluating antithrombotic and thrombolytic agents. In the present study we also evaluated the antithrombotic effects of vapiprost, a new thromboxane antagonist [2], and the thrombolytic activity of tissue plasminogen activator (tPA) using this model.

MATERIALS AND METHODS

Photochemically-induced thrombosis (PIT) model.

Male Wistar rats weighing 240 to 260 g were anesthetized by sodium pentobarbital (50 mg/kg, i.p.) and fixed on a heating-pad at 37 °C in the supine position. A 5 mm segment of the right femoral artery distal to the inguinal ligament was separated carefully and a pulse doppler flow probe (Model PDV-20 Crystal Biotech America, USA) was placed for monitoring blood flow. The contralateral femoral artery and vein were cannulated with polyethylene tubes for monitoring blood pressure, pulse rate and drug delivery, respectively. The arterial blood flow, blood pressure and pulse rate were continuously monitored on a thermal array recorder (WS-681G, Nihon-Kohden, Japan). Transillumination of green light (540 nm) was achieved by using a xenon lamp with a heat absorbing filter (Hamamatsu Photonics, Japan). The irradiation was led by an optic fiber (1 mm in diameter) positioned 5 mm away from the part of the right femoral artery proximal to the flow probe. This part of the artery was left intact. About 10 min after establishing the baseline blood flow, 10 mg/kg rose bengal (Wako, Japan) was injected, and then irradiation was started. Light exposure lasted 20 min. The femoral artery was considered occluded when the blood flow was completely stopped.

Key Words; arterial thrombosis, rat, photosensitization, vapiprost, tPA

Experiments with antithrombotic and thrombolytic agents

The antithrombotic effect of a new thromboxane antagonist, vapiprost (Glaxo), and the thrombolytic effect of tPA (TD-2061; Daiichi) were obtained as follows: Vapiprost (0.1, 0.3 or 1 mg/kg) was administered to the contralateral femoral vein 5 min before the injection of rose bengal, and in the other experiment, continuous infusion of tPA (30 or 100 µg/kg/min) was started 30 min after the establishment of thrombus by rose bengal injection under green light irradiation. The artery was judged to be reperfused when the blood flow was restored to more than 25% of baseline value.

Measurement of platelet aggregation

Blood was removed from the abdominal aorta 5 min after the injection of vapiprost, and deposited into a polypropylene tube containing 50 U/ml heparin (9:1 dilution). Platelet aggregation in whole blood was measured by impedance method using a Chrono-Log whole blood aggregometer (Chrono-Log, USA). Platelet aggregation in response to collagen (1 µg/ml; Hormon-Chemie, Germany) was studied for 10 min and the maximum aggregation was measured.

RESULTS

Photochemically-induced thrombus

Typical changes in the femoral arterial blood flow, blood pressure and pulse rate are shown in Fig.1. The blood flow of the irradiated femoral artery was completely occluded in 5.6 ± 0.34 min (mean ± SE, n=10) after the injection of 10 mg/kg of rose bengal with xenon light irradiation, without changes in blood pressure and pulse rate. Neither xenon light irradiation nor rose bengal injection could occlude the blood flow of the femoral artery without the concomitant use of them.

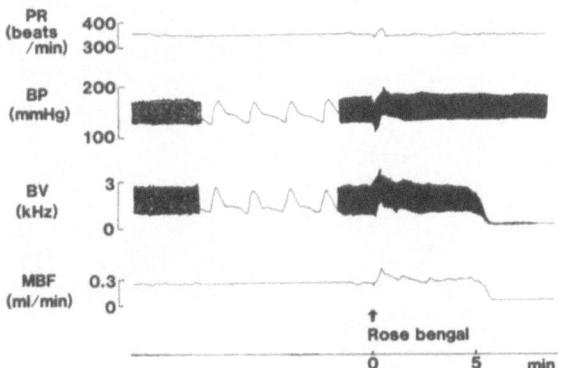

Fig.1 Records of typical changes in blood velocity (BV), in the mean blood flow (MBF) of the irradiated femoral artery and in pulse rate (PR) and blood pressure (BP) after the rose bengal (10 mg/kg) injection under xenon light irradiation.

The formation of massive thrombosis was evident by visual inspection. Transversal sections of the irradiated part of the femoral artery were examined by transmission electron microscope (TEM) (Fig.2). In contrast with the intact endothelium before rose bengal injection,

the endothelial layer was partly detached from the elastic membrane
3min after the injection (Panel A). After 5 min, platelets adherent
to or in close proximity to the endothelial surface were degranulated
and they possessed pseudopodia that extended towards the disrupted
endothelial layer (Panel B). Rouleaux of erythrocytes were observed
on the platelet clots. The longitudinal section of the irradiated
part of the femoral artery was examined by scanning electron
microscope (SEM). The luminal surface of the femoral artery was free
of adherent platelets before the rose bengal injection (Panel C).
After occlusion, the thrombus was containing many blood elements in
the lumen (Panel D). When the arterial segment was rinsed with
saline to remove the erythrocyte clots immediately after excision,
fibrin clots around platelets with pseudopods were observed at the
luminal surface (Panel E). After the establishment of reperfusion
which had been evoked by the administration of tPA (mentioned later),
few fibrin clots remained on the endothelial surface, but activated
platelets still adhered to it and the residual cells were still
swollen (Panel F).

Antithrombotic effect of vapiprost

As mentioned above, the femoral artery was completely occluded about
6 min after the rose bengal injection under green light irradiation.
Pretreatment with vapiprost (0.1, 0.3, and 1 mg/kg, i.v.) 5 min
before the injection of rose bengal prolonged the time required to
occlude the artery in a dose-dependent manner (Fig.3). In four of
six rats receiving 1 mg/kg vapiprost, complete cessation of blood
flow could not be obtained during the 60 min of observation, although
the blood flow itself was decreased. Ex vivo platelet aggregation of
whole blood in response to collagen was examined 5 min after the
injection of vapiprost. Vapiprost dose-dependently inhibited
platelet aggregation (Fig.3).

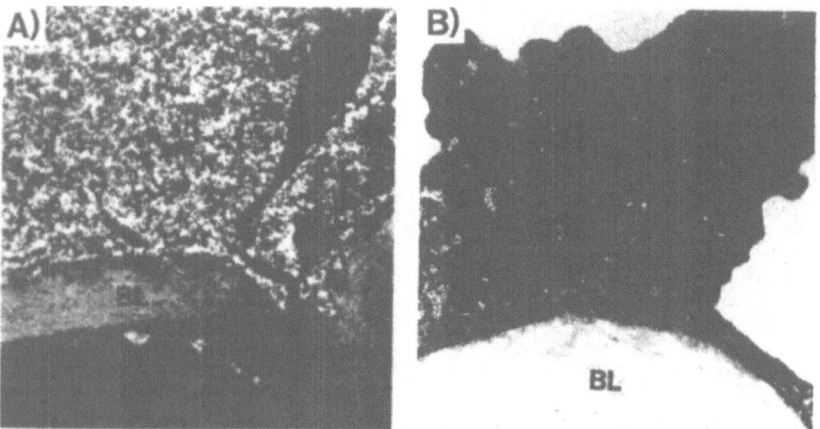

Fig.2 Observation of the irradiated segment of the femoral artery by
TEM (A,B) and SEM (C-F). Panel A: 3 min after the rose bengal
injection under green light irradiation, the endothelial cells were
detached from the basal lamina. Panel B: 5 min after the injection,
activated platelets adhered to the detached surface and accumulate at
that site.

25

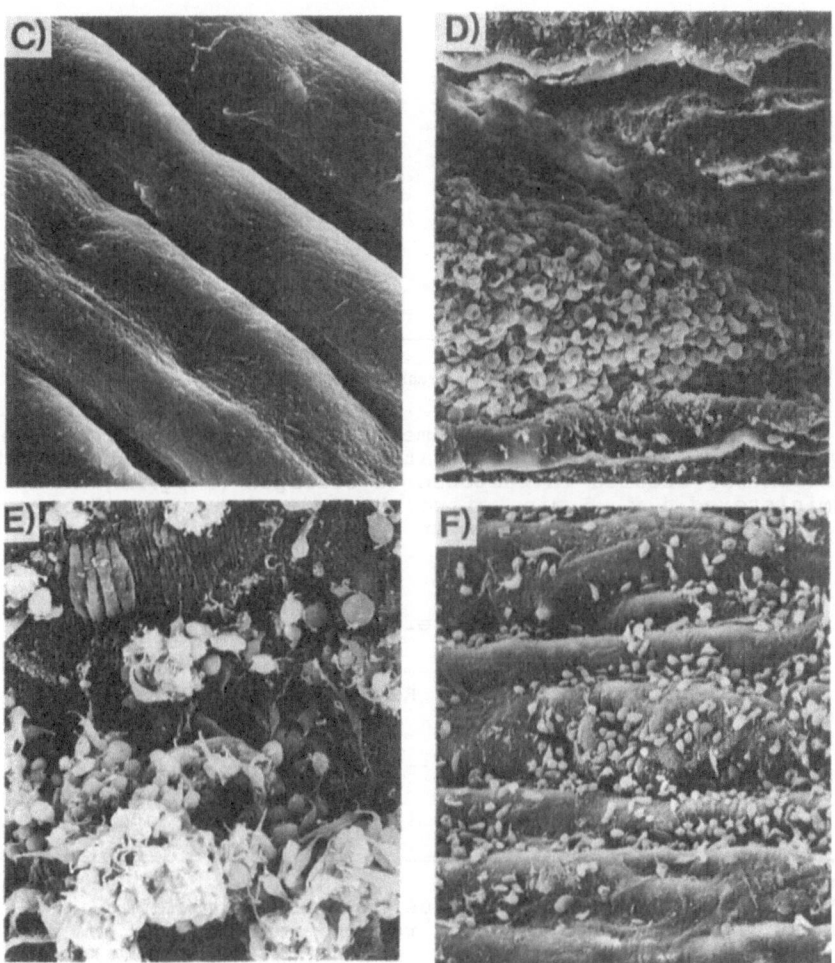

Fig.2 (continued) Panel C: Apparently intact luminal surface before
the injection.   Panel D: After occlusion, the lumen was packed with
red blood cells and other blood elements. Panel E: When the packed
lumen was rinsed with saline before fixation, fibrin clots and
platelets with pseudopods are shown to adhere to the luminal surface.
Panel F: After reperfusing with tPA in vivo, activated platelets with
pseudopods are still evident on the injured endothelial surface.

Thrombolytic effect of tPA

Spontaneous reperfusion of the occluded femoral artery did not recur
during the 150 min of observation after the complete interruption of
blood flow in the control rats given only saline.   In the rats
treated with tPA, reperfusion occurred in 2 of 10 rats given the
lower dose of 30 µg/kg/min, and in one of these, the reperfused
artery was immediately reoccluded.   In the higher dose group (100
µg/kg/min), 9 of 12 arteries were reperfused in 82 ± 2.9 min.
Reocclusion was observed in five of nine of the reperfused arteries
(Table 1).

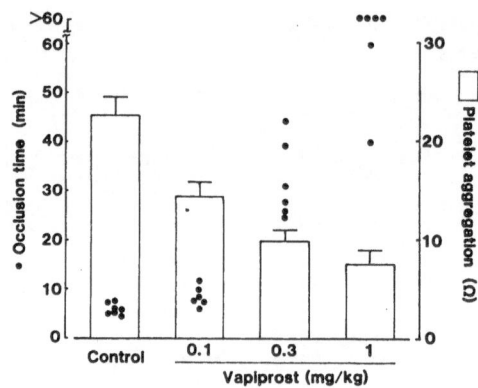

Fig.3 Effects of vapiprost on time required to occlude the femoral artery, and on platelet aggregation of whole blood. Platelet aggregation was studied in response to collagen (1 μg/ml). Vapiprost was administered i.v. 5 min before the injection of rose bengal (10 mg/kg). Each column indicates mean ± SE.

Table 1 Thrombolytic effect of tPA

Agent	Dose[a]	Reperfusion Incidence	Reperfusion Time (min)	Reocclusion Incidence
saline		0/12	–	–
tPA	30	2/10	(84, 136)[b]	1/2
	100	9/12	82 ± 2.9[c]	5/9

a:μg/kg/min, for 30 min.
b:Values are those that reperfusion could be observed.
C:Mean ± SE of 5 rats in which reperfusion could be observed.

DISCUSSION

The purpose of this study was to produce a simple and reproducible small animal model of arterial thrombosis for study of antithrombotic and thrombolytic agents. We applied the photochemical technique as a means of producing endothelial injury in the rat femoral artery. The fact that neither transillumination by xenon light irradiation nor rose bengal injection could occlude the blood flow of the femoral artery without the concomitant use of the other demonstrated the need for a photosensitized dye to initiate thrombus formation. Rose bengal is well known to be one of the most efficient photodynamic generators of molecular oxygen singlet [3]. This highly reactive form of oxygen may react with structural proteins and lipids in cellular membranes to initiate a sequence of direct peroxydation reactions leading to the endothelial damage [1,4]. Our histopathological observations suggested that the photochemical reaction was the initial event in the endothelial injury and that it was followed by activation of the platelets and eventual platelet rich thrombus formation.

To confirm that the model would be useful for evaluating the effect of antithrombotic agents, we evaluated the effect of a thromboxane antagonist, vapiprost, on thrombogenesis. Vapiprost is a potent anti-platelet drug [2], which was confirmed in the present study in rats: vapiprost inhibited ex vivo platelet aggregation of whole blood in a dose-dependent manner. Pretreatment with vapiprost dose-dependently prolonged the time required to occlude the artery, and we could find the correlation between the antithrombotic effect and the anti-platelet effect of vapiprost.

On the other hand, tPA is now used clinically as a thrombolytic agent because it has a specific affinity to plasminogen within the established thrombus and generates plasmin to cause fibrinolysis. In the present study the thrombolytic effect of intravenous tPA (30 and 100 µg/kg/min) could be demonstrated in a dose dependent manner, and by electron microscopic observation of the vessel after thrombolysis by tPA, many platelets were still seen to be intact, adhering to the endothelium. This finding is consistent with the fact that the reperfused vessel was frequently reoccluded afterwards.

In conclusion, the photochemically-induced thrombus (PIT) model in the rat femoral artery is quite simple and can induce arterial thrombus resembling the arterial thrombus in man [5] with high reproducibility. This model is expected to be a useful tool for evaluating antithrombotic and thrombolytic agents.

REFERENCE

1. Watson BD, Dietrich WD, Busto R, Wachtel MS, Ginsberg MD (1985) Ann Neurol 17: 497-504

2. Lumley P, White BP, Humphrey PPA (1989) Br J Pharmacol 97: 783-794

3. Gandin E, Lion Y, Van de Vorst (1983) Photochem Photobiol 37: 271-278

4. Herrmann KS (1983) Microvasc Res 26: 238-249

5. Friedman MF, Van der Boverkamp EJ (1966) Am J Pathol 48: 19-44

CONVERSION OF FIBRINOGEN TO FIBRIN BY THROMBIN AND ADSORPTION OF THROMBIN BY FIBRIN CLOTS -- THE OBSERVATIONS AND A HYPOTHESIS*

CHUNG YUAN LIU, DANNY LING WANG, AND YU-TAI PAN

Institute of Biomedical Sciences, Academia Sinica, Taipei, Taiwan, Republic of China.

INTRODUCTION

Thrombin (EC 3.4.21.5), a serine protease, generated by proteolytic cleavage of prothrombin (Factor II) through the concerted action of several enzymes and cofactors (1) has many effects on the hemostatic system including conversion of fibrinogen to fibrin (Fig. 1), activation of Factors V, VIII, XIII, and protein C, and stimulation of platelet aggregation and release. Early studies on thrombin reaction with fibrinogen showed decreased thrombin concentration in the clot supernatant (2). The amount of thrombin lost was later found to be directly proportional to the concentration of fibrinogen (3), and also depended upon the initial concentration of thrombin (4). Kinetic studies (5) indicated that fibrin formation followed pseudo-first order kinetics with respect to fibrinogen.

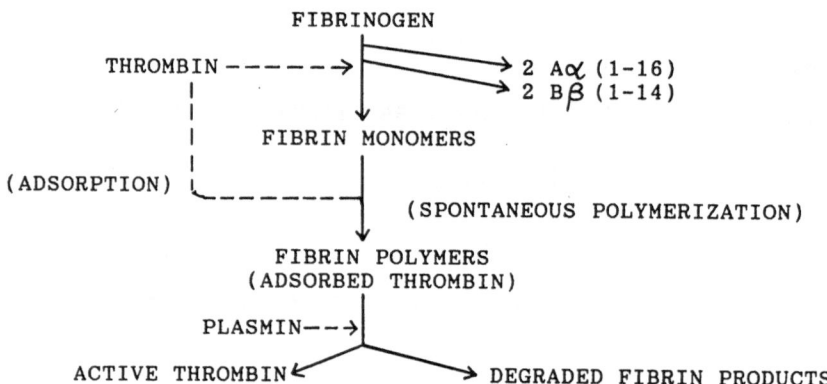

Figure 1. Thrombin interaction with fibrinogen and fibrin.

---

* The experimental work was completed mainly in Columbia University College of Physicians & Surgeons, New York, NY, USA.

Clinical reports (6, 7) of thrombosis occurring in patients with thromboembolic disorders following the treatment with streptokinase or urokinase suggest that the dissolution of fibrin can be associated with new thrombosis. In order to account for these observations, we assume that thrombin can be adsorbed by fibrin clots and that adsorbed thrombin is released from the dissolved fibrin clots and prompted new thrombosis. The present report is designed to define the nature of the interaction between thrombin and fibrin both in a purified system and the application of these observations in human blood and plasma as a basis for later studies on the possible physiological importance of the reaction.

MATERIALS AND METHODS

Materials and methods were as previously described elsewhere by Liu, et. al. (8, 9, 10, 11).

RESULTS AND DISCUSSION

Conversion of fibrinogen to fibrin by thrombin and adsorption of thrombin by fibrin clots:

Human fibrinogen can be converted to fibrin monomer by thrombin with the release of two fibrinopeptides A [Aα (1-16)] and two fibrinopeptides B [Bβ (1-14)], and fibrin monomers formed can spontaneously polymerize to form fibrin polymers or fibrin clots with the adsorption of thrombin (Fig. 1). Thrombin can be adsorbed by fibrin clots as free components in the fluid phase within the fibrin clots or as bound components to fibrin strands of the clots. Thrombin adsorbed remains active but the adsorption physically localizes and limits thrombin action. Also the adsorption can prevent thrombin action with its inhibitors.

Specific binding of thrombin by fibrin:

The specific binding of thrombin to fibrin has been studied. Fig.2A shows that thrombin binding by fibrin increases with increasing thrombin concentration and is virtually identical whether measured as clotting activity or as radioactivity, demonstrating that the labeled and unlabeled thrombin binding is in the same manner. When the results are analyzed by a Scatchard plot (Fig. 2B), nonlinearity is apparent but the curve can be approximately resolved into two

Table I

Summary of apparent association constants and maximum molar ratios of thrombin binding to fibrin

Material and method	Max. molar binding ratio n, (Thrombin/Fibrin)	Association Constant Ka, (1/M)
Human system		
High affinity binding	0.39	5.8×10^5
Low affinity binding	1.6	6.8×10^4
Bovine system		
Scatchard plot	0.47	4.0×10^5
Steck-Wallach plot	0.46	4.0×10^5
Sips plot	0.5	3.9×10^5

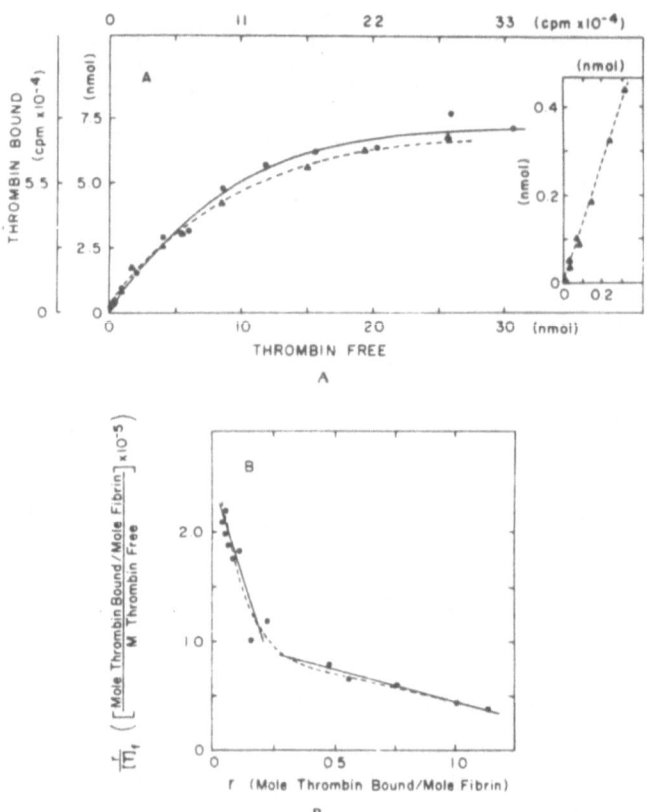

Figure 2. Binding of thrombin to fibrin as a function of thrombin concentration. For detailed experimental procedures see reference (8). (A) Increasing amounts of thrombin were added to separate tubes with constant amounts of fibrinogen (2 mg) in barbital buffer at 25 °C, final reaction volume 0.7 ml. Non-cross-linked fibrin clots were formed: (●——●) thrombin by clotting activity; (▲--▲) ^{125}I-thrombin by radioactivity. (B) A Scatchard plot of ^{125}I-thrombin binding by fibrin. The plot was similar when thrombin clotting activity was measured.

straight lines, suggesting two types of binding site. A summary of the maximum molar binding ratio of thrombin to fibrin and the association constant of the thrombin-fibrin binding estimated from this plot are shown in Table I. These studies indicate the existence of equilibrium binding between thrombin and fibrin. Such a mechanism is compatible with product inhibition of thrombin by fibrin, consistent with the observation that fibrin acts like an antithrombin. Thrombin adsorbed by fibrin clots can be released by plasmin digestion of the fibrin clots. These observations suggest a physiological significance for these reactions. The results of ultrastructural studies of thrombin-gold conjugate binding by fibrin clots (Fig. 3) are consistent with the results described above. At a concentration of 0.1 μM thrombin, 73% of the observed conjugates are bound at branch points of the fibrin network (Fig. 3A, B). At higher thrombin concentrations (e.g., 0.5 μM) only 38% of the conjugates are bound at branch points while 62% are bound within the linear portion of the fibrin strand (Fig. 3C, D), suggesting that the binding is specific and that there may be at least two different affinity sites.

Analyses of the bovine system (10) indicate that only a single class of binding sites is suggested, and that the results appear to resemble those for the higher affinity binding site in the human system (Table I). These analyses have been based on the results from the Scatchard plot, the Steck-Wallach plot, and the Sips plot (10).

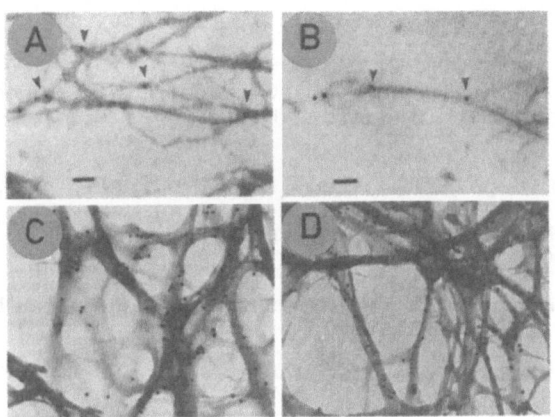

Figure 3. Electron micrographs showing the structural interaction of thrombin-gold (T-Au) conjugates with fibrin strands. (A) and (B) at a concentration of 0.10 μM, the T-Au appearing as an integral part of the fibrin strand at branch points (arrow heads) when viewed by negative staining (A) and by thin sectioning after resin embedment (B). (C) and (D) T-Au concentration of 0.50 μM (negative staining) before (C) and after (D) 72-hours dialysis of the clot against 0.15 M NaCl, pH 7.0. For detailed procedures see reference (11).

A HYPOTHESIS

Fibrin can instantaneously and reversibly adsorb thrombin, and can further immobilize thrombin by fibrin polymerization, limiting and localizing thrombin action to the location where it is generated. On the other hand, antithrombin III and heparin can strongly bind and inactive thrombin in blood circulation. Thrombin adsorbed can be released by plasmin proteolysis of the fibrin clots.

It is intriguing to speculate that the reactions of fibrin together with those of antithrombin III and heparin may be regarded as a cooperative defense mechanism against excessive thrombin activity generated in the circulation as a response to injury. Accordingly, blood clotting is not only a primary defense mechanism against hemorrhagic trauma but also a well-organized defense mechanism against excessive clotting activity in circulation. In addition, the adsorption of thrombin by fibrin also provides the thrombin source required for factor XIII activation in the fibrin clots.

REFERENCES

1. Suttie,J.W., and Jackson,C.M. (1977) Physiol. Rev. <u>57</u>, 1-70.

2. Eagle,H. (1935) J. Gen. Physiol. <u>18</u>, 547-555.

3. Wilson,S. (1942) Arch. Intern. Med. <u>69</u>, 647-661.

4. Seegers,W.., Nieft,M., and Loomis,E.C. (1945) Science <u>101</u>, 520-521.

5. Waugh,D.F., and Livingstone,B.J. (1951) J. Phys. Colloid Chem. <u>55</u>, 1206-1218.

6. Sashara,A.A., Hyers,T., Cole,C., Ederer,F., Murray,I.A., Wenger, N., Sherry,S., and Stengle,J.M. (1973) Circulation <u>47</u> (Suppl.II), 1-108).

7. A National Cooperative Study (1974) J.A.M.A. <u>229</u>, 1006-1613.

8. Liu,C.Y., Nossel,H.L., and Kaplan,K.L. (1979) J. Biol. Chem. <u>254</u>, 10421-10425.

9. Liu,C.Y., Kaplan,K.L., Markowitz,A.H., and Nossel,H.L. (1980) J. Biol. Chem. <u>255</u>, 7627-7630.

10. Liu,C.Y. (1980) Enzyme <u>25</u>, 64-68.

11. Liu,C.Y., Handley,D.A., and Chien,S. (1985) Analytical Biochem. <u>147</u>, 49-56.

LOW MOLECULAR WEIGHT HEPARIN IN THE TREATMENT OF DEEP VEIN THROMBOSIS OF INFERIOR LIMBS

STANISLAW LOPACIUK, for the Polish Venous Thrombosis Study Group.
Institute of Hematology, Warsaw, Poland

INTRODUCTION

Unfractionated heparin (UFH) is still the most commonly administered drug in the treatment of deep vein thrombosis (DVT). It may be given intravenously either by continuous infusion or by intermittent injection. Of these routes, continuous infusion is preferred because it produces less bleeding complications. UFH may also be given by subcutaneous injection twice daily. Recent clinical trials heve shown that subcutaneous UFH is a safe and effective alternative to continuous intravenous UFH in the treatment of DVT (1).

It has recently been found that low molecular weight (LMW) heparins exhibit pharmacokinetic properties suitable for sub-cutaneous treatment (2). Clinical trials have demonstrated that these agents are effective in the prevention of DVT following general (3, 4, 5) and orthopedic surgery (6). Reports on the use of LMW heparins for the treatment of established DVT are less numerous. Of the published randomized trials, only three studies (7, 8, 9) have been sufficiently large to yield significant results. In a study performed by Bratt et al. (7), a LMW heparin (Fragmin, Kabi) was administered subcutaneously in doses adjusted to plasma anti-Xa activity and found to be as effective as continuous intravenous UFH. In studies of Ninet and Duroux (8) and Prandoni et al. (9), patients with DVT were randomized to either subcutaneous Fraxiparine (CY 216), a LMW heparin from Choay, in a fixed dose or continuous intravenous UFH. These two trials showed that subcutaneous Fraxiparine is more effective than UFH in terms of phlebographic resolution of thrombi after 10 days of treatment. The present trial was performed to determine

Key words: deep vein thrombosis, low molecular weight heparin, unfractionated heparin

the efficacy and safety of subcutaneous Fraxiparine compared with subcutaneous UFH as the initial treatment for DVT of the lower limbs.

PATIENTS AND METHODS

The study was designed as a prospective, open, stratified and randomized trial with a blind evaluation of phlebographic results. Six Polish centers participated in the trial, the inclusion period was 22 months, from February 1989 to December 1990.

Patients of either sex over 16 years of age with phlebographically proved proximal or calf DVT and duration of symptoms not longer than 10 days were enrolled in this study. Exclusion criteria were: clinically suspected pulmonary embolism, phlegmasia caerulea dolens, treatment with heparin or oral anticoagulant prior to admission, recent history of DVT (less than 2 years), surgery or trauma within the previous 3 days, contraindication to heparin therapy, pregnancy, and documented antithrombin III deficiency.

Patients were stratified into two subgroups on the basis of whether they had a proximal or calf DVT (with or without involvement of the popliteal vein), then they were randomly allocated by using a sealed envelope to either the Fraxiparine or UFH group. Patients in the Fraxiparine group were given subcutaneous injections of this drug (CY 216, Choay) in a fixed dose of 225 anti-Xa ICU/kg every 12 h. In the UFH group the treatment was initiated by an intravenous bolus dose of 5000 IU of UFH (Calciparine, Choay). The drug was then administered by subcutaneous injection every 12 h; for the first 2 or 3 injections the dose was calculated as 250 IU/kg/12 h, the subsequent dosage was adjusted daily to maintain the activated partial thromboplastin time (APTT) between 1.5 and 2.5 times patient's basal value.

The treatment with Fraxiparine or UFH lasted 10 days. Oral anticoagulation with acenocoumarol (Sintrom, Geigy) was introduced on the 7th day of heparin administration and continued for at least three months in both groups. The trial protocol was approved by the ethical committee of the Institute of Hematology, Warsaw and all patients gave their written consent to participate in the study.

Ascending phlebography was carried out before entry into the trial and repeated on the first or second day after the end of Fraxiparine or UFH treatment. Phlebograms were assessed blindly by an independent panel of three experts using the score system of Arnesen et al. (10). Since many films did not permit definite assessment of the iliac vein, this vessel wase excluded from quantitation. The maximal possible score for involvement of the deep veins was 25 points.

All patients were examined daily during the first 10 days of therapy; symptoms of recurrent DVT, pulmonary embolism or bleeding were sought. The patients were then followed for three months to monitor oral anticoagulation and to assess the possibility of thrombotic recurrence. Hematocrit estimations and urine analyses were carried out daily during the whole period of Fraxiparine or UFH administration, and platelet counts were estimated on days 2, 5, 7 and 10. Blood samples for APTT estimations were obtained before the morning injections of Fraxiparine or UFH.

Statistical analysis was performed by the Fisher exact test, Student t-test and chi-square test. A probability value less than 0.05 was considered significant.

RESULTS

A total of 149 consecutive patients were entered into trial. After randomization and treatment, 3 patients in the UFH group were judged to be truly inelegible and excluded from all analyses. Of the remaining 146 patients, 74 were treated with Fraxiparine and 72 received UFH. The treatment groups did not differ in basic features such as age, weight, sex, extension of thrombi, duration of symptoms, history of previous DVT incidence of factors predisposing to DVT.

The efficacy of treatment was assessed on the basis of phlebographic findings in 134 patients. The mean Arnesen's score before treatment was similar in the Fraxiparine and UFH patients (15.6 ±0.8 and 16.8 ±0.8, respectively). After 10 days of treatment, in both groups the mean score decreased significantly in comparison to the baseline one (to 13.5 ±0.7 in the Fraxiparine patients and to 15.0±0.8 in the UFH patients, p<0.001). However, the difference in score changes between the two groups was not significant (p>0.05). There was phlebographic improvement in 45/68 patients (66%) in the Fraxiparine group and in 32/66 patients (48%) in the UFH group, and increase in the thrombus size in 10/68 (15%) and 12/66 (18%), respectively. The difference in the thrombus fate between the two groups did not reach statistical significance (p = 0.05).

One symptomatic non-fatal pulmonary embolism and one major hemorrhage were observed in the UFH group. During a follow-up period of 3 months, two rethromboses occured in the UFH group and none in the Fraxiparine group. One patient died of renal failure (in the course of systemic lupus erythematosus) two months after the treatment with UFH.

Mean hematocrit values in the Fraxiparine and UFH patients were similar both at entry and on the 10th day of treatment. A slight fall in the platelet count (to 70-90 x $10^9/l$) was fonud in three patients during the therapy with UFH. The APTT was maintained within the therapeutic range in a least 80% of patients during the administration of UFH. In the Fraxiparine group, the APTT remained practically unchanged in comparison to the pretreatment values.

CONCLUSIONS

Subcutaneous fixed dose Fraxiparine is safe and at least as effective as adjusted unfractionated heparin in the initial treatment of deep vein thrombosis. Since the administration of Fraxiparine does not require laboratory monitoring, it has practical and economic adventages over standard heparin therapy.

REFERENCES

1. Hommes DW, Bura A, Mazzolai L, Buller HR, ten Cate JW (1991) Thromb Haemostas 65: 753.
2. Fareed J, Walenga JM, Rancanelli A, Hoppenstead D, Coyne E, Davis P, Nonn R (1989) In: Fraxiparine, Breddin K, Fareed J, Samama M (eds) Schattauer. Stuttgart-New York, pp. 41-67.
3. Kakkar VV, Djazaeri B, Fok J, Fletcher M, Scully MF, Westwick J (1982) Br Med J 284: 375-379.
4. Caen JP (1988) Thromb Haemostas 59: 216-220.
5. The European Fraxiparin Study (EFC) Group (1988) Br J Surg 75: 1059-1063.

6. Turpie AGG, Levine M. N., Hirsh J, Carter CJ, Jay RM, Powers PJ, Andrew M, Hill RD, Gent M (1986) N Engl J Med 315: 925-929.

7. Bratt G, Aberg W, Johansson M, Tornebohm, Granqvist, Lockner D (1990) Thromb Haemostas 64: 506-510.

8. Ninet J, Duroux P (1990) In: Fraxiparine, Bounameaux H, Samama M, ten Cate JW (eds) Schattauer. Stuttgart-New York, pp. 63-75.

9. Prandoni P, Vigo M, Cattelan AM, Ruol A (1990) Haemostasis 20 (suppl 1): 220-223.

10. Arnesen H, Heilo A, Jakobsen E, Ly B, Skaga E (1978) Acta Med Scand 203: 457-463.

THE MECHANISM OF BLOOD ANTICOAGULATION BY LEECH HEMENTIN

ANDREI Z. BUDZYNSKI

Department of Biochemistry, Temple University School of Medicine,
Philadelphia, PA 19140, U.S.A.

SUMMARY

Anticoagulation of blood by the leech *Hirudo medicinalis* is accomplished by hirudin, a potent inhibitor of thrombin. The giant South American leech *Haementeria ghilianii* also renders blood incoagulable in order to feed and digest the meal. However, the mechanism of blood anticoagulation by the latter leech is different from the former. The anticoagulant, called hementin, is present in both the anterior and posterior salivary glands. It is a metalloproteinase that specifically cleaves fibrinogen rendering it incoagulable by thrombin. The mechanism of degradation by this enzyme involves preferential hydrolysis of three peptide bonds in the coiled-coil connector, resulting in destruction of bivalent function of the fibrinogen molecule. Since the other coagulation factors in human blood are unaffected by hementin, the mechanism of action of this anticoagulant is selectively focused on fibrinogen.

INTRODUCTION

The discovery of blood anticoagulant in leeches in 1884 by John Haycraft [1] followed by thorough studies of Fritz Markwardt and colleagues resulted in biochemical characterization of hirudin [2] and its clinical utilization as an anti-thrombotic agent [3]. Originally, hirudin has been purified from the medicinal leech, *Hirudo medicinalis*. Several hirudin forms have been obtained recently by the recombinant DNA technology. The anticoagulant effect of hirudin on blood is mediated through extraordinary high affinity for thrombin (Figure 1). Hirudin forms with thrombin an inactive enzyme-inhibitor complex in which both the catalytic site and fibrinogen binding anionic exosite are blocked [4].

A distant relative of the medicinal leech is the giant leech, *Haementeria ghilianii*, native to French Guyana and neighboring Amazonia. The full grown leech is approximately 50 cm in length, providing a lot of tissue for experiments. Another advantage of this leech is the ability to grow and reproduce in captivity on a large scale. Both *Hirudo medicinalis* and *Haementeria ghilianii* live on mammalian blood, however, the two species belong to different taxonomic orders. It was conceivable, therefore, that the two leeches may have developed different mechanisms of interference with blood coagulation in order to maintain the blood in a liquid state for feeding and digestion. Research on the nature of blood anticoagulant
from *Haementeria ghilianii* has been focused on the salivary glands as the source of the material. Studies summarized below demonstrated that the blood anticoagulant was entirely different from hirudin. The anticoagulant has been named hementin by onomatopoetic relation to the word hirudin.

RESULTS AND DISCUSSION

Two pairs of the salivary glands, anterior and posterior, are present in the leech *Haementeria ghilianii*. The anterior gland has a large size and is composed of large cells while the posterior gland has a small size and is composed of small cells [5]. Electrophoretic analysis of the salivary gland homogenate revealed entirely different protein composition of the two glands [6]. This observation raised the issue of whether the same forms of the anticoagulant occur in the two salivary glands. Many experiments were performed to document whether isozymes of hementin are present in the leech, however, a clear answer to this problem has not been obtained as yet.

Hementin represents the main anticoagulant of the salivary glands that occurs only in the cytosol [6]. The anterior gland contains low concentration of hementin while the posterior gland has high concentration. The difference in the size of the glands makes it so that the total amount of hementin per either anterior or posterior gland is about the same. Hementin is a proteinase that rapidly degrades fibrinogen and fibrin, of either human or bovine origin, in purified systems and in plasma [7]. Degradation of the fibrinogen is not affected by calcium ions regardless of whether the anterior or posterior gland extract has been tested. Hementin appears to be the predominant proteinase in these extracts since upon electrophoretic

ANTICOAGULANTS FROM LEECHES

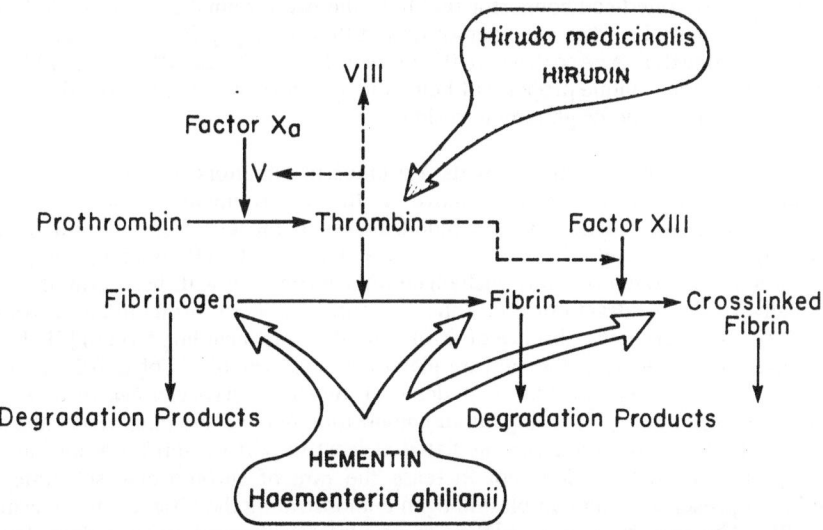

Figure 1.   Mechanism of action of anticoagulants from leeches. The main target for hirudin is thrombin and inactivation of this enzyme arrests progression of blood coagulation. In the liquid blood, hementin primarily attacks fibrinogen so no clot can be formed because of lack of the substrate. In the clotted blood, hementin also will degrade fibrin.

separation under non-denaturing conditions only one zone of lysis has been observed using human non-cross-linked fibrin as a substrate. However, a very small difference between electrophoretic mobilities of hementin derived from the anterior and posterior salivary gland extracts would support a possible existence of hementin isozymes with different charges [7]. Another evident factor favoring hementin isozymes came from observations that the rate of fibrinogen degradation by the anterior salivary gland extract was slightly faster than that by the posterior gland extract. Moreover, anterior gland hementin seemed to generate more degraded

and more heterogeneous fibrinogen fragments than posterior gland hementin [8].
Similar patterns of fibrinogen degradation have been observed using anterior and posterior salivary gland extracts from the leech *Haementeria officinalis* [8] implying that hementin-type anticoagulant may be a characteristic component in *Haementeria* species. However, the individual leeches in this taxonomic order may not necessarily have the same functional form of hementin.

The mechanism of blood anticoagulation by hementin involves degradation of fibrinogen. The Michaelis constant for human fibrinogen is 1 μM. The depletion of the coagulable substrate allows the leech to keep the blood in a liquid form even in the presence of thrombin (Figure 1). Fibrin is also susceptible to degradation by this enzyme but non-cross-linked clots are lysed at a faster rate than cross-linked clots. The cleavage of purified carboxymethylated polypeptide chains of human fibrinogen is very slow, implying that the native conformation of the substrate may play a role in the recognition mechanism [9]. The pattern of fibrinogen degradation by hementin resembles that caused by plasmin since products analogous to plasmic fragments Y, D and E are generated [10,11] that have been characterized and designated as Y_{hem}, D_{hem}, and E_{hem} [11,12]. Hementin has a unique low affinity for the COOH-terminal domain of the α-chain, in contrast to most proteolytic enzymes which preferentially cleave fibrinogen in this domain. Hementin from the posterior gland has even less affinity for the α-chain COOH-terminus than that from the anterior gland [8,12]. Thus, hementin is a unique enzyme that spares proteinase-sensitive α-chain COOH-terminus and cleaves the fibrinogen molecule preferentially in the coiled-coil connector that links the two terminal domains with the central one. Detailed structural studies have demonstrated that three peptide bonds were cleaved in the coiled-coil connector: AαAsn102-Asn103, BβLys130-Gln131, and γPro76-Asn77 [11]. The finding showed another unique property of hementin, which cleaved peptide bonds at the NH_2-terminal side of asparagine or glutamine residues.

Another unique property of hementin is its resistance to inhibitors of proteinases present in human blood. The enzyme remains active in this environment for several hours [7]. Fibrinogen is the main substrate in the blood, however, the other coagulation factors are unaffected so that the level of factors II, V, VII, IX, X, XI, XII, prekallikrein and high molecular weight kininogen remains unchanged after incubation with hementin at 37° for 30 min [3]. Salivary gland extract does not inhibit reaction between thrombin and chromogenic substrate S-2238 indicating the absence of thrombin inhibitors including hirudin [7]. However, other inhibitors were found in the salivary gland extracts. Inhibitors of trypsin, plasmin, α-chymotrypsin and granulocyte elastase were demonstrated in leech species *Haementeria ghilianii* and *Haementeria officinalis* [13]. Lung tumor colonization in mice was inhibited by the salivary gland extracts [14]; this response was mediated at least in part by antistasin, an inhibitor of factor Xa [15]. The extract does not increase the rate of chromogenic substrate S-2251 cleavage in the presence of human plasminogen demonstrating the absence of a plasminogen activator [7]. The extract inhibits ADP-induced platelet aggregation in a dose-dependent manner. Since the initial aggregation rate is the same in the absence and the presence of the extract, and since a disaggregation of platelets occurs after prolonged incubation with the extract, the inhibitory effect is ascribed to the degradation of fibrinogen by hementin but not to the presence of a putative disintegrin [8].
Purification of hementin from the salivary gland extracts is very difficult because of instability of the enzyme and irreversible inactivation. Many metal chelating buffers, including Tris, rapidly destroy activity of hementin. Using series of inhibitors against main
classes of proteolytic enzymes, it has been inferred that hementin appears to be a calcium-regulated zinc-containing metalloproteinase. The enzyme purified by ammonium sulfate precipitation, ion-exchange chromatography and preparative polyacrylamide gel electrophoresis in non-denaturing conditions had a molecular weight of 120,000 [9]. However, purification by differential ultrafiltration, ion-exchange chromatography and reverse-phase HPLC gave a preparation with a molecular weight of 80,000 [16].

ACKNOWLEDGEMENTS

The author would like to thank Dr. Elizabeth Lasz for her help and critical comments during the preparation of the manuscript. This work was supported in part by Grant HL36221 from the National Heart, Lung and Blood Institute, National Institutes of Health, Bethesda, MD, USA.

REFERENCES

1. Haycraft JB. (1884) Ueber die Einwirkung eines Sekretes des officinellen Blutegels auf die Gerinnbarkeit des Blutes. Arch Exp Pathol Pharmakol 18: 209-217.
2. Markwardt F. (1988) Biochemistry and pharmacology of hirudin. In: Pirkle H, Markland Jr FS, eds. Hemostasis and animal venoms. Marcel Dekker, New York, Basel, pp 255-269.
3. Markwardt F. (1991) The comeback of hirudin as an antithrombotic agent. Seminars in Thrombosis and Hemostasis 17:79-82.
4. Rydel TJ, Ravichandran KG, Tulinsky A, Bode W, Huber R, Roitsch C, Fenton II JW. (1990) The structure of a complex of a recombinant hirudin and human α-thrombin. Science 249:277-280.
5. Sawyer, RT, Damas, D, Tomic MT. (1982) Anatomy and histochemistry of the salivary complex of the giant leech *Haementeria ghilianii* (Hirudinea: Rhynchobdellida). Arch Zool Exp Gen 122:411-425.
6. Budzynski AZ, Olexa SA, Sawyer RT. (1981) Composition of salivary gland extracts from the leech *Haementeria ghilianii*. Proc Soc Exp Biol Med 168:259-265.
7. Budzynski AZ, Olexa, SA, Brizuela BS, Sawyer RT, Stent GS. (1981) Anticoagulant and fibrinolytic properties of salivary proteins from the leech *Haementeria ghilianii*. Proc Soc Exp Biol Med 168:266-275.
8. Budzynski AZ. (1991) Interaction of hementin with fibrinogen and fibrin. Blood Coagulation and Fibrinolysis 2:149-152.
9. Malinconico SM, Katz JB, Budzynski AZ. (1984) Hementin: anticoagulant protease from the salivary gland of the leech *Haementeria ghilianii*. J Lab Clin Med 103:44-60.
10. Malinconico SM, Katz JB, Budzynski AZ. (1984) Fibrinogen degradation by hementin, a fibrinogenolytic anticoagulant from the salivary glands of the leech *Haementeria ghilianii*. J Lab Clin Med 104:842-854.
11. Kirschbaum NE, Budzynski AZ. (1990) A unique proteolytic fragment of human fibrinogen containing the Aα COOH-terminal domain of the native molecule. J Biol Chem 265:13669-13676.
12. Kirschbaum NE, Budzynski AZ. (1988) Purification and characterization of a fibrinogen fragment containing the COOH-termini of all three chains of the native molecule. In: Mosesson MW, Amrani DL, Siebenlist KR, DiOrio JP, eds. Fibrinogen 3. Biochemistry, biological functions, gene regulation and expression. Excerpta Medica, Amsterdam, pp. 297-300.
13. Murer EH, James HL, Budzynski AZ, Malinconico SM, Gasic GJ. (1984) Protease inhibitors in *Haementeria* leech species. Thromb Haemostas 51:24-26.
14. Gasic GJ, Viner ED, Budzynski AZ, Gasic GP. (1983) Inhibition of lung tumor colonization by leech salivary gland extracts from *Haementeria ghilianii*. Cancer Res 43:1633-1636.
15. Tuszynski GP, Gasic TB, Gasic GJ. (1987) Isolation and characterization of antistasin an inhibitor of metastasis and coagulation. J Biol Chem 262:9718-9723.
16. Swadesh JK, Huang I-Y, Budzynski AZ. (1990) Purification and characterization of hementin, a fibrinogenolytic protease from the leech *Haementeria ghilianii*. J Chromatog 502:359-369.

REGULATORY MECHANISMS OF THROMBOMODULIN EXPRESSION IN THE HUMAN UMBILICAL VEIN ENDOTHELIAL CELLS

S. OHDAMA, S. TAKANO, S. MIYAKE, K. HIROKAWA AND N. AOKI

The First Department of Internal Medicine, Tokyo Medical and Dental University, Tokyo, Japan.

INTRODUCTION

Thrombomodulin (TM) is a cell surface glycoprotein that plays an important role in inhibition of intravascular coagulation as a cofactor for the thrombin-catalyzed activation of protein C.[1] It contributes to thromboresistant properties of endothelium. We reported that TM-fragments increased in plasma of patients with various thrombotic disorders such as DIC, ARDS and pulmonary thromboembolism.[2] However, it is not clear whether the increase of TM-fragments in plasma is caused simply by vascular damages or reflects an increase of expression of TM in endothelial cells. In this paper, we summarize our studies on the regulatory mechanism of TM expression in vitro assuming its role in pathogenesis of hyper-coagulable states.

MATERIALS AND METHODS

Cell culture Human umbilical vein endothelial cells (HUVECs) were isolated from umbilical cord veins by treatment with 0.2 % collagenase and grown to confluent in α-MEM supplemented with 20 % FCS, 5 u/ml heparin, 10 μg/ml basic FGF. 100 u/ml penicillin G and 100 μg/ml streptomycin. The cells were subcultured by brief exposure to 0.25 % trypsin and 0.04 % EDTA and harvested in 35 mm diameter plastic petri dishes coated previously with gelatin. All experiments were performed with confluent endothelial cell monolayers from the second to fifth passage.

Assay for cell surface TM antigen Surface TM antigen was determined using HUVEC monolayers ($1.0-1.5 \times 10^5$ cells/well) in 24-well tissue culture plates. The cells were incubated with or without agents for the indicated times at 37°C, 5 % CO_2. After incubation, the treated cells were washed twice with α-MEM containing 0.5 % bovine serum albumin. Then, aliquots of 200 μl of horseradish peroxidase conjugated

42

monoclonal anti-TM antibody KA₂ (0.1 µg/ml),[3] were added to the wells and wells were incubated for 15 min at 37°C. After washing in the same way, 0.01 % H_2O_2 and 0.8 mg/ml ortho-phenylene diamine dissolved in 0.075 M citrate-sodium phosphate buffer pH 5.0, were placed in each well and incubated for 30 min at 37°C. The reaction was stopped by addition of 50 µl of 4.5 M H_2SO_4 to each well, and the absorbance at 492nm was measured by a spectrophotometer.

<u>Cyclic</u> <u>AMP</u> <u>Measurement</u> Cyclic AMP was measured by a radioimmunoassay using Cyclic AMP Kits Yamasa and results were expressed as picomoles of cAMP per mg of protein.

<u>Northern</u> <u>blotting</u> Total cellular RNA was extracted from confluent monolayers by treatment with guanidium isothiocyanate and centrifugation through cesium chloride. The RNA was transferred to a HyBond-N filter by standard capillary blotting techniques. The filter was hybridized overnight with ^{32}P-labelled complementary DNA. Autoradiography was performed using Kodak XAR film. After densitometric analysis of autoradiographs, TM mRNA levels were normalized to the concentration of β-actin mRNA.

RESULTS AND DISCUSSION

<u>Effects</u> <u>of</u> <u>cytokines</u> <u>and</u> <u>endotoxin</u> <u>on</u> <u>TM</u> <u>expression</u> <u>of</u> <u>HUVECs</u> We found that tumor necrosis factor (TNF), interleukin-1 (IL-1) and lipopolysaccharide (LPS) suppress surface TM expression,[4] as previously reported.[5-8] The reduction of surface TM antigen was reported to be associated with increased endocytosis and lysosomal degradation of TM. These cytokines also inhibited transcription of TM mRNA, and subsequent decrease in TM protein synthesis may also contribute to decline in TM expression.

<u>Effects</u> <u>of</u> <u>agents</u> <u>that</u> <u>can</u> <u>increase</u> <u>intracellular</u> <u>cAMP</u> Surface TM antigen was increased by agents that can increase intracellular cAMP[4,9]; pentoxifylline (PTX), theophylline (TP), dibutyryl cAMP (dbcAMP) and forskolin. Up-regulation of TM by PTX and dbcAMP was due to de novo synthesis of TM protein resulting from increased TM mRNA levels. In addition, suppression of surface TM antigen by cytokines was abrogated by PTX and dbcAMP. These results suggested proein kinase A may be involved in up-regulation of TM expression and may protect partially against cytokine-induced endothelial cell injury and restore anticoagulant state of endothelium. Dibutyryl

cGMP had no effect on TM activity, so protein kinase G may not be involved in cellular regulatory mechanism for TM expression.[10]

Effects of phorbol esters Effects of phorbol myristate acetate (PMA) were biphasic, with reduced TM activity on cell surfaces occurring between 1-6 h, followed by enhancement at 24 h.[4,10] Chloroquinine increased total TM antigen in cell lysates of HUVECs incubated with PMA, suggesting that the inhibition of TM activity involved internalization and degradation of TM.[10] Dittman et al. reported that PMA had no effect on TM mRNA levels in mouse hemangioma cells. Scarpari and Sadler also reported that PMA had no effect on TM mRNA in HUVECs. But our observation demonstrated that PMA increased TM mRNA. Differing results among these reports may come from differences in experimental conditions, species, normal cells and transformed cells, and number of passage. These results suggested that protein kinase C may be involved in up- and down-regulation of TM expression.

Effect of vasoactive amines Histamine increased TM activity and TM mRNA by activation of H_1-receptors on the cells, but serotonin and bradykinin had no effect.[11] Aspirin or indomethacin did not abrogate the enhancement of TM activity by histamine. These results suggested that the enhanced TM activity by histamine is not due to PGI_2 or other prostanoids produced by cycloxygenase.[11]

Effect of glucocorticoid Dexamethasone (DEX) significantly increased surface TM expression by inducing up-regulation of TM mRNA (unpublished observation).[12]

SUMMARY

TM expression in endothelial cells is modulated by mechanisms involving TM gene transcription and presumably by internalization / degradation of TM molecules. Expression of TM in HUVECs is regulated by various agents: endotoxin and cytokines (TNF, IL-1) down-regulate TM; forskolin and cAMP analogs up-regulate TM; PMA have biphasic effects, down-regulation for a short period (1-6 h) followed by up-regulation; histamine up-regulate TM by activation of H_1 receptor; dexamethasone up-regulate involving cytoplasmic TM mRNA. These results suggested that TM expression on endothelial cells plays a role in regulation of intravascular coagulation in various disease states, as its possible control by pharmacological means.

REFFERENCES

1. Esmon C.T. and Owen W.G. (1981) Identification of an endothelial cell cofactor for thrombin-catalyzed activation of protein C. Pro. Natl. Acad. Sci. U.S.A. 78:2249-2252.

2. Takano S., Kimura S., Ohdama S. and Aoki N. (1990) Plasma thrombomodulin in health and diseases. Blood 76:2024-2029.

3. Kimura S., Nagoya T. and Aoki N. (1989) Monoclonal antibodies to human thrombomodulin whose binding is calcium dependent. J. Biochem. 105:478-483.

4. Hirokawa K. and Aoki N. (1990) Up-regulation of thrombomodulin in human umbilical vein endothelial cells in vivo. J. Biochem. 108:839-845.

5. Nawroth P.P. and Stern D.M. (1986) Modulation of endothelial cell hemostatic properties by tumor necrosis factor. J. Exp. Med. 163:740-745.

6. Conway E.M. and Rosenberg R.D. (1988) Tumor necrosis factor suppresses transcription of the thrombomodulin gene in endothelial cells. Mol. Cell. Biol. 8:5588-5592.

7. Nawroth P.P., Handley D.A., Esmon C.T. and Stern D.M. (1986) Interleukin 1 induces endothelial cell procoagulant while suppresssing cell-surface anticoagulant activity. Proc. Natl. Acad. Sci. U.S.A. 83:3460-3464.

8. Moore K.L., Andreoli S.P., Esmon N.L., Esmon C.T. and Bang N.U. (1987) Endotoxin enhances tissue factor and suppresses thrombomodulin expression of human vascular endothelium in vitro. J. Clin. Invest. 79:124-130.

9. Ohdama S., Takano S., Miyake S. and Aoki N. (1991) Pentoxifylline prevents tumor necrosis factor induced suppression of endothelial cell surface thrombomodulin. Thromb. Res. 62:745-755.

10. Hirokawa K. and Aoki N. (1991) Regulatory mechanisms for thrombomodulin expression in human umbilical vein endothelial cells in vitro. J. Cell. Physiol. 147:157-165.

11. Hirokawa K. and Aoki N. (1991) Up-regulation of thrombomodulin by activation of histamine H_1-receptors in human umbilical vein endothelial cells in vitro. Biochem. J. 276:739-743.

12. Ohdama S. and Aoki N. in preparation.

LOCALIZATION OF HEPARIN-LIKE COMPOUNDS IN CULTURED AORTIC ENDOTHELIAL CELLS

K. TAKEUCHI, K. SHIMADA[1], M. NISHINAGA, S. KIMURA, T. OZAWA
Department of Medicine and Geriatrics, Kochi Medical School, Kochi 783, and
[1]Department of Cardiology, Jichi Medical School, Tochigi 329-04, Japan

INTRODUCTION

The non-thrombogenic properties of blood vessels are due, in part, to anticoagulant heparin-like compounds on the luminal surface of vascular endothelial cells. A certain class of heparan sulfate proteoglycans (HSPGs) associated with the endothelial surface can bind antithrombin III (AT III) in plasma and can accelerate the thrombin inactivation by the protease inhibitor. HSPGs are also known to be located along the abluminal side of the endothelium, namely basement membrane or extracellular matrix (ECM) produced by endothelial cells[1]. Thus, questions exist regarding the localization of anticoagulant HSPG in endothelial cells. Does ECM contain HSPGs which interact with antithrombin III? If they exist, how abundant are they, as compared to those in the luminal surface?

METHODS

ECM was prepared from porcine aortic endothelial cell cultures by removing the cells with nonenzymatic methods as previously described[2]. HSPG in ECM was labeled metabolically by incubating the cells with ^{35}S-sulfate prior to the preparation of ECM[3]. Antithrombin III binding to ECM was performed by incubating ECM with ^{125}I-labeled antithrombin III for 30 minutes at 37°C, followed by rapid washes of dishes to remove free antithrombin III[4]. Heparin-like activity of ECM was demonstrated by the acceleration of thrombin inactivation by antithrombin III. Residual thrombin activity after incubation of the enzyme with antithrombin III on the ECM-coated dishes was measured by chromogenic substrate[5]. HSPGs were isolated from ECM and subjected to the antithrombin III affinity chromatography. The HSPGs with high affinity for antithrombin III were eluted with the higher concentration of NaCl[4].

RESULTS

The time course of AT III binding to ECM was shown in Fig. 1. Nonspecific binding was defined as the amount of radiolabeled AT III bound in the presence of excess concentration of unlabeled AT III. Specific binding was then obtained by subtracting nonspecific binding from total binding. The binding reached equilibrium by 30 minutes of incubation. The AT III binding to ECM was then performed with increasing concentrations of ^{125}I-labeled antithrombin III, and compared with the binding to intact endothelial cells. AT III bindings to both intact cell and ECM were saturable at approximately 100 nM of AT III, but the maximum binding to ECM was approximately 40% of that in the intact cell (Fig. 2). Fig. 3 shows the displacement curve of the AT III bound to ECM by various GAGs. Heparan sulfate was more potent in the displacing ability than chondroitin sulfates or dermatan sulfates. In Fig. 4, the ECM-coated dishes were treated with GAGlyases prior to the AT III binding experiment. As compared to the control, after treatments with heparitinase, the binding was almost completely abolished. On the other hand, after treatments with chondroitinase ABC or AC, the reduction of the binding was only minimal. Thus, removal of heparan sulfate, but not chondroitin sulfate or dermatan sulfate, destroyed the AT III binding ability of ECM.

In the following series of experiments, ^{35}S-labeled proteoglycans in ECM was analyzed. ^{35}S-radioactivity incorporated into ECM-associated GAGs was approximately 40% of that in the intact cell, indicating that at least 40% of total GAGs in the intact cell may be located in the ECM (Fig. 5). The experiment was then performed to identify the proportion of different GAGs in ^{35}S-prelabeled ECM. After treatment with heparitinase, ^{35}S-GAGs in ECM was less than 10% of that in control ECM. On the other hand, chondroitinase treatments resulted in only a

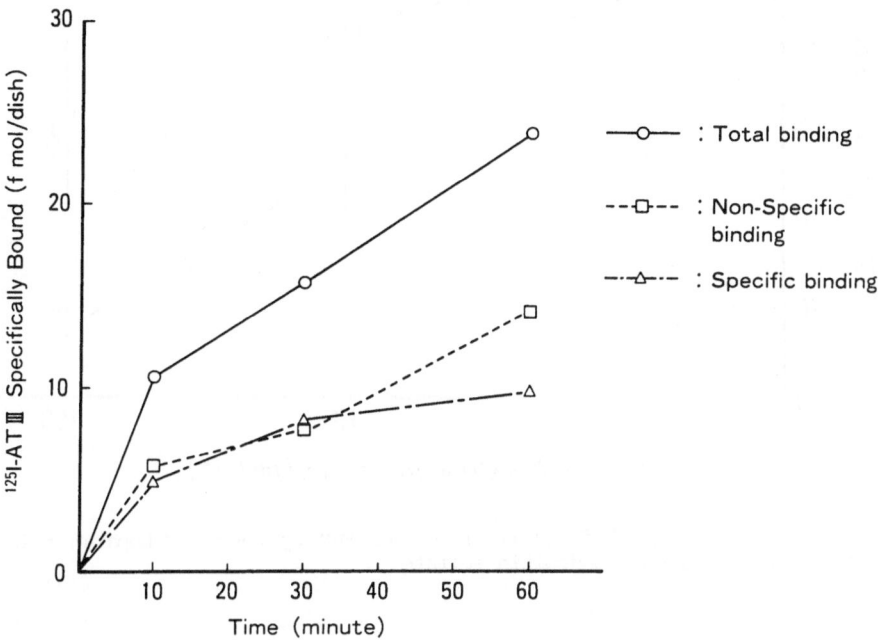

Fig. 1. Time course of AT III binding to ECM.

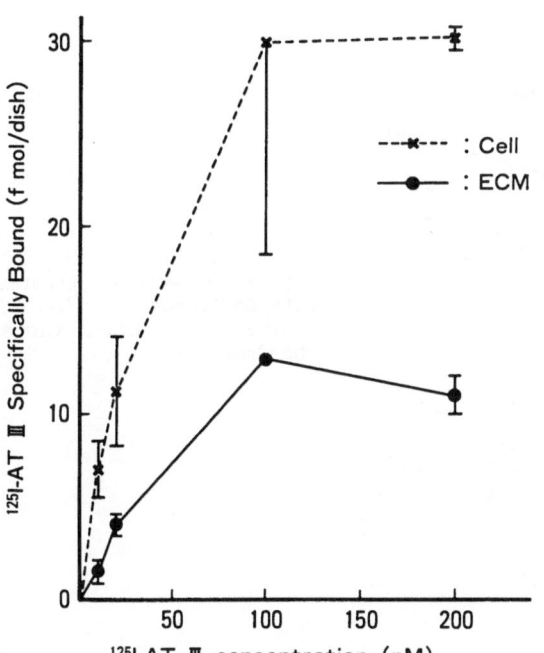

Fig. 2. Dose-dependency of AT III binding to endothelial cell and ECM

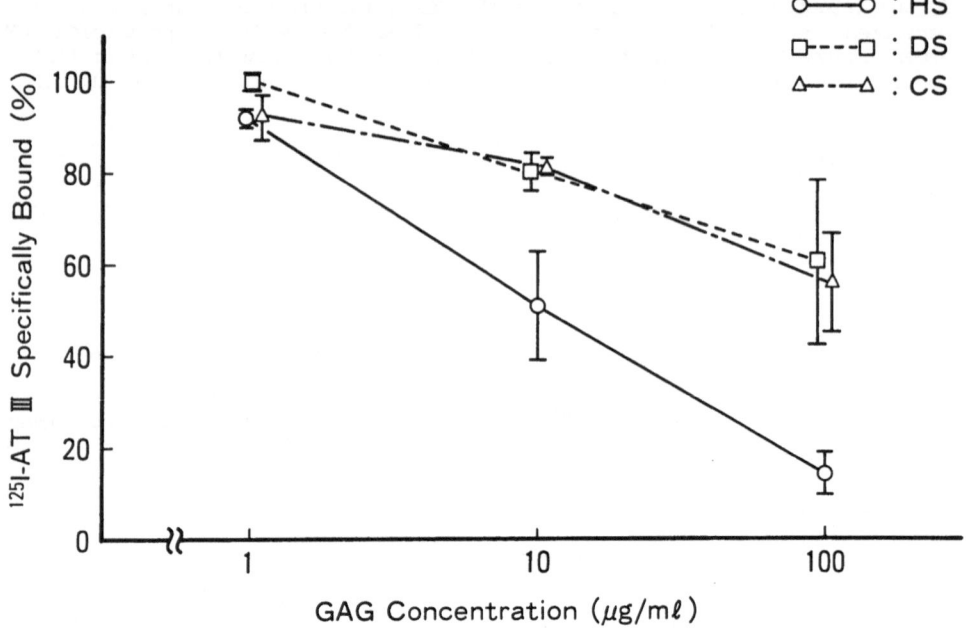

Fig. 3. Displacement of ^{125}I-AT III bound to ECM by GAGs. HS:heparan sulfate, DS:dermatan sulfate, CS:chondroitin sulfate

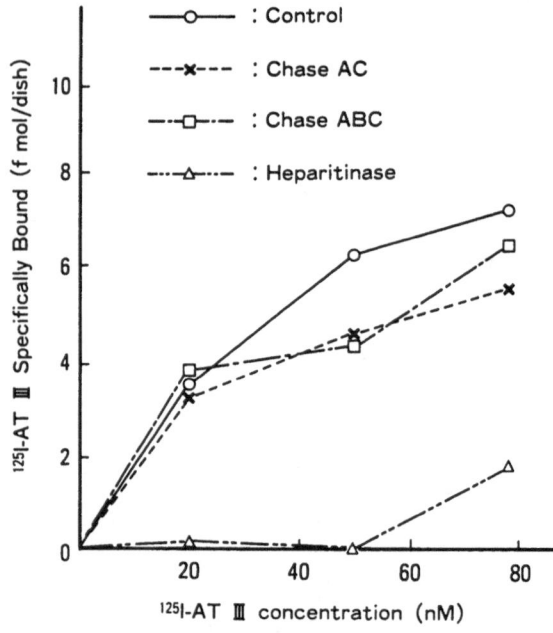

Fig. 4. Effects of treatment with GAGlyases on ^{125}I-AT III binding to ECM. Chase: chondroitinase.

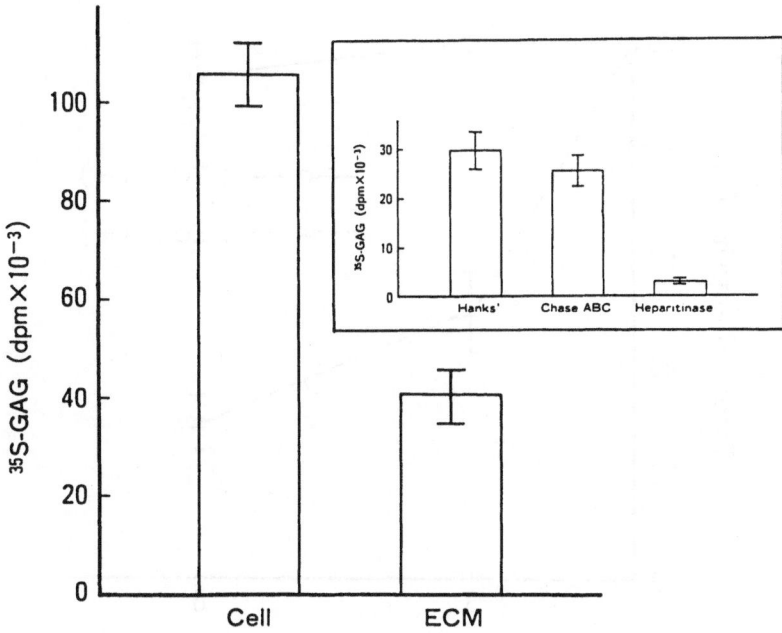

Fig. 5. The radioactivity incorporated into total cell- and ECM-associated ^{35}S-GAG. Inset: Effects of treatment with GAGlyases on ECM-associated ^{35}S-GAG

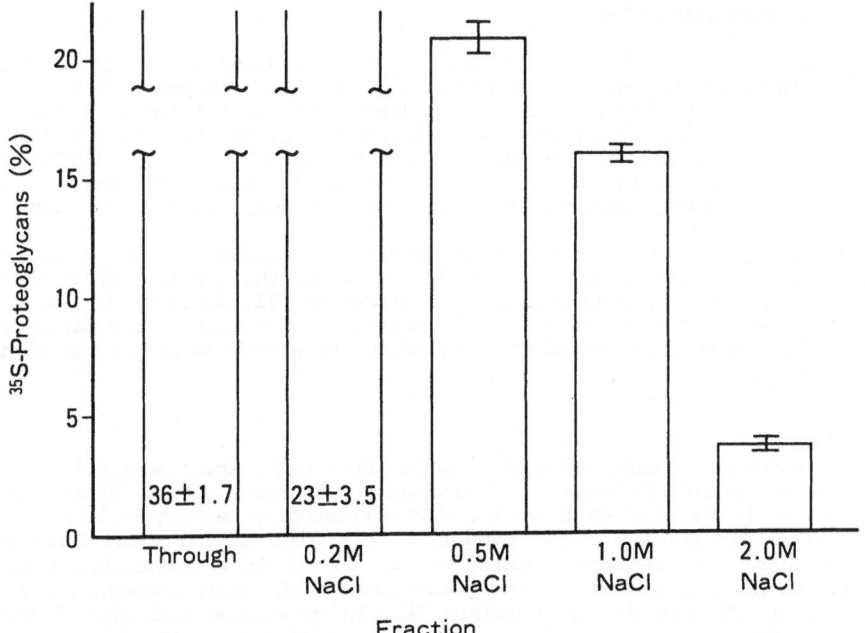

Fig. 6. AT III affinity chromatography of ECM-associated ^{35}S-proteoglycans.

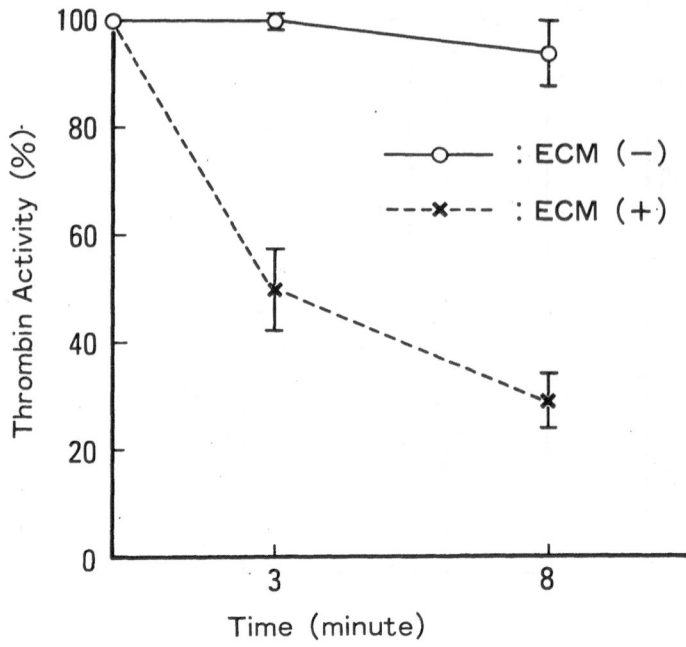

Fig. 7. Heparin-like activity of ECM. Thrombin was incubated with antithrombin III for the indicated period of time. Residual thrombin activity is plotted against incubation time.

minimal reduction of ^{35}S-GAGs, indicating that heparan sulfate is the major constituent of the endothelial ECM-associated GAGs. ^{35}S-proteoglycans were then isolated from prelabeled ECM and applied onto the antithrombin III affinity column. Fig. 6 shows that the major portion was eluted in a through fraction or with lower concentration of NaCl. Approximately 15 to 20% of HSPGs, however, were eluted with more than 1 M concentration of NaCl. This indicates that ECM contains, in fact, heparin-like compounds with high affinity for antithrombin III.

The experiment was then performed to examine whether ECM can accelerate the thrombin inactivation by antithrombin III. In the absence of ECM, thrombin activity was slowly inactivated by antithrombin III. As shown in Fig. 7, residual thrombin activity was minimally reduced by 8 minutes of incubation time. On the ECM, however, the thrombin inactivation by antithrombin III was significantly accelerated.

DISCUSSION

ECM produced by cultured aortic endothelial cells specifically binds AT III. The amount of AT III bound to ECM accounted for approximately 40% of that in the intact cell. The binding is efficiently displaced by heparan sulfate, and almost completely abolished by pretreatment of ECM with heparitinase. ECM-associated HSPGs apparently represent approximately 40% of HSPGs associated with intact cell, in parallel with the binding experiments. ECM does contain, in part, HSPG with high affinity for antithrombin III. In consistent with this finding, ECM accelerates inactivation of thrombin by antithrombin III. We conclude that approximately at least a half of anticoagulant HSPG is located in the ECM, that is, abluminal side of cultured aortic endothelial cells.

Our data are in accordance with the previous study[6] of de Agostini et al. in which they showed that the most of antithrombin III bound to the endothelium is localized immediately beneath the aortic endothelium in vivo. Only a very small extent of labeled antithrombin III was noted by autoradiography on the luminal surface of the endothelial cells. It is, however, difficult to estimate the exact amount or proportion of antithrombin III bound to the basement membrane by the histochemical technique they used. Furthermore, the technique we used in the present study does not allow us to know whether all of the ECM beneath the cultured endothelial cells remained intact on petri dishes. Thus, although the data suggest that at least a half of HSPGs appears to be localized in the ECM, the precise quantitative analysis of localization still remains to be determined.

Heparin-like activity of ECM may represent a "reservoir" that could be brought into play with substantial damage to the overlying endothelium. Alternatively, it may well function in situ, as plasma antithrombin III should have relatively free access to this site[7], thereby constituting the anticoagulant activity within subendothelial regions. In any case, subendothelial extracellular matrix may provide another critical biological surface in contact with the hemostatic mechanism.

REFERENCES
1. Pejler G., Backstrom G., Lindahl U., Paulsson M., Dziadek M., Fujiwara S. and Timpl R. (1987) Structure and affinity for antithrombin of heparan sulfate chains derived from basement membrane proteoglycans. J. Biol. Chem. 262:5036-5043.

2. Shimada K., and Ozawa T. (1987) Subendothelial extracellular-matrix heparan sulfate proteoglycan-degrading activity of human monocyte macrophages. Heart Vessels 3:175-181.

3. Shimada K., and Ozawa T. (1985) Evidence that cell surface heparan sulfate is involved in the high affinity thrombin binding to cultured porcine aortic endothelial cells. J. Clin. Invest. 75 :1308-1316.

4. Shimada K., and Ozawa T. (1987) Modulation of glycosaminoglycan production and antithrombin III binding by cultured aortic endothelial cells treated with 4-methylumbelliferyl- β -D-xyloside. Arteriosclerosis. 7 :627-636.

5. Shimada K., Kawamoto A., Matsubayashi K., and Ozawa T. (1989) Histidine-rich glycoprotein dose not interfere with interactions between antithrombin III and heparin-like compounds on vascular endothelial cells. Blood. 73 :191-193.

6. de Agostini A.I., Watkins S.C., Slayter H.S., Youssoufian H. and Rosenberg R.D. (1990) Localization and anticoagulantly active heparan sulfate proteoglycans in vascular endothelium: Antithrombin binding on cultured endothelial cells and perfused rat aorta. J. Cell Biol. 111:1293-1304.

7. Carlson T.H., Simon T.L. and Atencio, A.C. (1985) In vivo behavior of human radioiodinated antithrombin III: distribution between three physiologic pools. Blood 66:13-19.

THE EFFECT OF HUMAN GRANULOCYTIC NEUTRAL PROTEASES ON HEMOSTASIS: EFFECTS OF CATHEPSIN G ON PLATELETS

MARIA KOPEC AND KSENIA BYKOWSKA
Laboratory of Blood Coagulation and Hemostasis,
Institute of Hematology, Warsaw, Poland.

INTRODUCTION

Human neutrophil elastase (HNE) and cathepsin G (Cat G) are serine proteases active at neutral pH (1, 2, 3) and both are released from polymorphonuclear (PMN) azurophilic granules during phagocytosis, blood coagulation, strenuous exercise, under the influence of immune complexes, chemotactic peptides and sodium urate crystals (4).

HNE and cat G contribute to the essential PMN functions of bacteriolysis, in initiating of inflamation and tissue injury (5, 6). These proteinases can also interfere with the function of all main components of the hemostatic system i.e. coagulation, fibrynolysis, blood vessels and platelets. HNE and cat G have been shown to degrade and inactivate numerous factors of blood coagulation and fibrynolysis as well as their ihibitors (7). HNE was reported to detach endothelial cells and to digest the intracellular matrix (5,8). Cat G was described to release PGI_2 from vascular endothelium but to suppress the thrombin mediated enhancement of PGI_2 production (8).

Cat G and thrombin differ distinctly in their effects on platelets (9). The former aggregates platelets and provokes the serotonin release (10). The latter does not induce the release reaction and its aggregating potential is very low (9,11). Exhaustive proteolysis by HNE as well as by cat G abolishes platelet responses to thrombin and to ristocetin plus vWF most probably due to degradation of their receptor PGIb on platelets (10, 12, 13).

The aim of the present study was to compare the effects of cat G on human platelets with those mediated by thrombin.

Key words: Leukocytic proteases, platelets.

MATERIALS AND METHODS

Materials were purchased from the following sources: thrombin Biomed (Lublin, Poland), alfa-chymotrypsin, EGTA and ADP Sigma (St. Louis, USA), N-benzoyl-L-tyrosine ethyl ester (BTEE), leupeptin (Ac-Leu-Leu-Arginal) and neomycin Calbiochem (Lucerne, Switzerland), N-ethylmaleimid (NEM) Serva (Heidelberg, DBR), PGI_2, Epoprostenol Wellcome Research Labs (Beckenham, UK). Iloprost was kindly offered by dr. E. Schillinger (Schering AG, Berlin DBR). Cat G was purified according to Baugh and Travis (14). It was dissolved in 0.05 M Tris buffer containing 0.5 M NaCl (pH 6.8) and stored at − 20°C before use.

Suspensions of human platelets were prepared and their treatment with enzymes performed as described earlier (10).

Platelet aggregation was examined in Chronolog aggregometer at 37°C.

^{3}H-serotonin release was determined as described by Bykowska et al (9).

PF4 release was measured using Egzygnost PF4 kit from Behring (BDR).

Binding of ^{125}I-fibrinogen to platelets was determined according to Cierniewski et al (15).

RESULTS

The potency of cat G to expose platelet receptors for ^{125}I-fibrinogen appeared high (112,000/platelet), higher than that of ADP (20,000) and close to that of α − chymotrypsin (128,000).

Cat G aggregated platelets and induced ^{3}H-serotonin release (60,4%) as efficiently as thrombin (61,6%).

Leupeptin inhibited platelet aggregation by thrombin in a dose dependent fashion and abolished it completely at a final concentration of 1.25 mg/ml. In contrast, cat G provoked aggregation was only slightly suppressed by leupeptin 200 mg/ml. The enzymatic cat G activity on BTEE was also not influenced by leupeptin.

EGTA 2mM and indomethacin added up to a concentration of 40mM did suppress neither thrombin nor cat G induced platelet aggregation. NEM 0.016 mM reduced both by about 75%.

Much more resistant was the platelet aggregation by cat G than that provoked by thrombin to the inhibitory action of PGI_2 and

54

iloprost. PGI₂ at final concentration of 10 ng/ml abolished entirely the thrombin effect while cat G mediated aggregation was not
completely inhibited by 40 ng/ml of PGI₂. (Fig. 1).

Even more striking was the difference in the effect of iloprost.
Platelet aggregation by thrombin was completely suppressed by 2
ng/ml but its 1000 fold higher concentration inhibited only partially cat G aggregatory action.

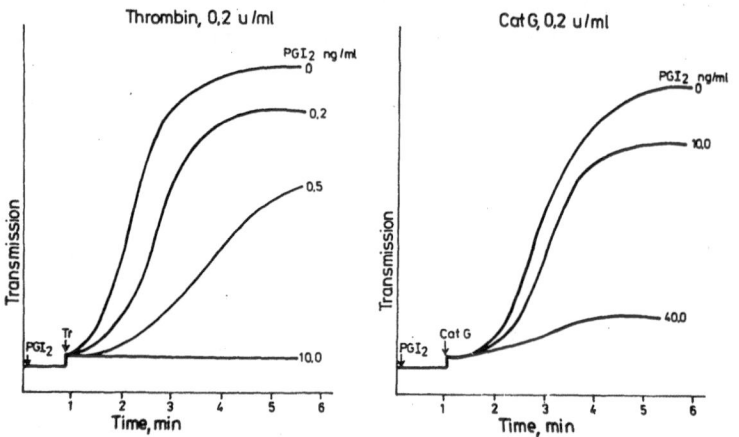

Fig 1. Effects of PGI₂ on platelet aggregation.

Cat G induced PF4 secretion appeared also much more resistant
to iloprost than that mediated by thrombin. Iloprost in
a concentration of 2ng/ml suppressed completely PF4 release but
even 200ng/ml of iloprost induced only partial (64%) inhibition.

Pronouncendly less sensitive was also the aggregatory potential
of cat G than that of thrombin to neomycin. Its 2mM concentration
induced complete inhibition of aggregation by thrombin but only
10% reduction in cat G effect.

DISCUSSION

The results of our study indicate that cat G is a strong
platelet agonist which exsposes a high number of fibrinogen

receptors on platelets, induces the aggregation and releases platelet dense and α -granule constituents. The potency of cat G to aggregate human platelets and to induce the release reaction is similar to that of thrombin. The responses of platelets to cat G and to thrombin are both resistant to Ca^{2+} chelation by EGTA and to the inhibition of cyclooxygenase pathway of the arachidonic acid metabolism by indomethacin. Effects of both cat G and thrombin are suppressed by NEM to similar extent.

Distinctly different if compared with thrombin appeared, however, the susceptibility of cat G mediated platelet effects to same other inhibitors. Leupeptin known to inhibit thrombin amidolytic activity and its effects on platelets (16) influenced neither enzymatic cat G activity on synthetic substrate nor platelet aggregation by this protease. PGI_2 and its stable analogue iloprost, both recognised to suppress platelet activation due to intracellular cAMP increase (17), appeared much less potent inhibitors of cat G induced platelet aggregation and release reaction if compared with responses to thrombin. Two to three orders magnitude higher iloprost concentrations were required to obtain similar degrees of inhibition of cat G action.

Pronouncendly less susceptible was also aggregation by cat G than by thrombin to neomycin which is considered to inhibit the thrombin provoked activation of phosphoinositol production in platelets (18). Mechanisms underlying the differences in platelet responses to thrombin and cat G remain to be elucidated. Different may be pathways of signal transmission from the receptors of these enzymes. Inactivation by cat G of its own receptor should also be taken into account and it is subject of our present study.

REFERENCES
1. Starkey PM. (1977) In: Barret AJ. (ED) Proteinases in mammlian cells and tissues. North Holland Publishing Company. Amsterdam, New York, Oxford. pp 57-89.
2. Janoff A. (1986) Physiol Rev 36: 207-216
3. Travis J. Salvesen GS. (1983) Ann Rev Biochem 52: 655-709
4. Wachtfogel YT. Abrams W. Kucich U. Weinbaum, Schapira H. Colman RW. (1988) J Clin Invest 81: 1310-1316
5. Harlan JM. (1987) Semin Thromb Haemost 13: 434-444

6. Weiss JS. (1989) N Engl J Med 320: 365-376

7. Machovich R. Owen WG. (1990) Blood Coagul Fibrynol 1: 79-90

8. Weksler BB. Jaffe EA. Brower MS. Cole OF. (1989) Blood 74: 1627-1634

9. Bykowska K. Kaczanowska J. Karpowicz M. Stachurska J. Kopeć M. (1983) Thromb Haemost 50: 768-772

10. Bykowska K. Kaczanowska J. Karpowicz M. Łopaciuk S. Kopeć M. (1985) Thromb Res 38: 538-546

11. Kornecki E. Ehrlich YH. De Mars DD. Lenox RH. (1986) J Clin Invest 77: 750-756

12. Brower MS. Levin R. Garry K. (1985) J Clin Invest 75: 657-666

13. Wicki AN. Clemetson KJ. (1985) Eur J Biochem 153: 1-11

14. Baugh RJ. Travis J. (1976) Biochemistry 15: 836-841

15. Cierniewski CS. Kowalska MA. Krajewski T. Janiak A. (1982) Bioch Biophys Acta 714: 543-548

16. Ruggiero M. Lapetina EG. (1986) Thromb Res 42: 247-255

17. Lerea KM. Glomset JA. Krebs EG. (1987) J Biol Chem 262: 282-288

18. Siess W. Lapetina EG. (1986) FEBS Lett 207: 53-57

Platelets

GLYCOPROTEIN IIb/IIIa AND MEMBRANE SKELETON IN RESTING AND ACTIVATED PLATELETS

H. SUZUKI,[1] K. TANOUE[1] AND H. YAMAZAKI[1,2]

[1]Department of Cardiovascular Research, The Tokyo Metropolitan Institute of Medical Science, Tokyo 113 and [2]Jissen Women's University, Hino 191, Japan

INTRODUCTION

When platelets are activated by an appropriate stimulus, they change their shapes from discs to spheres with extending pseudopodia, release the contents of their storage granules, and aggregate with each other. These responses are believed to be mediated by the cytoskeleton, especially as a result of the contractile activity of actin filaments and associated proteins.[1,2] The human platelet membrane skeleton has been identified as a membrane-bound actin submembrane structure and that is distinct from the network of cytoplasmic actin filaments.[3] In the present study, we examined the cytoskeletal organization in unstimulated and thrombin-activated platelets, focusing on the relationship between membrane GPIIb/IIIa complex, a receptor for fibrinogen, and the membrane skeleton.

MATERIALS AND METHODS

Preparation of platelets and electron microscopy   Washed human platelets were obtained from the platelet rich plasma following Mustard's method as modified by Suzuki et al.[4]   The twice-washed platelets were finally suspended in Tyrode's solution (pH 7.4) containing 0.35% bovine serum albumin (BSA) and 0.1% glucose without added Ca^{2+} to a concentration of 5×10^5 platelets/μl.   The suspension of platelets with or without addition of thrombin (0.05 U/ml, final concentration) was added with equal volume of 0.04% dithiobis succinimidyl propionate (DTSP)[5] and incubated for 10 min at room temperature.   The DTSP-treated platelets were simultaneously extracted and fixed by mixing 1:1 with a solution containing 2% glutaraldehyde, 0.2% Triton X-100, 10 mM EGTA and 100 mM lysine, 100 mM sodium cacodylate, pH 6.8, according to the method of Boyles et al.[6]   After the addition of the solution, the tubes were inverted once and immediately centrifuged at 3,000 g for 2 min at 4°C.   The platelet pellets were rinsed with 0.1 M cacodylate buffer (pH 7.4) five times to remove Triton X-100, fixed again with 2% glutaraldehyde in the same buffer at 4°C for 30 min.   After rinsing five times with 0.1 M cacodylate buffer, the pellets were post-fixed with 1% osmium tetroxide in the same buffer for 60 min, rinsed with distilled water five times, immersed in 2% uranyl acetate in distilled water for 16 hr, dehydrated with a graded ethanol series, and embedded in Epon. Thin sections were stained on the grid with uranyl acetate and lead citrate and examined with a electron microscope at an accelerating voltage of 80 kV.

Immunocytochemistry   To determine the localization of GPIIb/IIIa, the unstimulated and thrombin-activated platelets were fixed by the addition of an equal volume of a solution containing 4% paraformaldehyde and 0.2% glutaraldehyde in 0.1 M phosphate buffer (pH 7.4) and kept for 30 min at room temperature.   The fixed platelets were centrifuged at 3,000 g for 2 min.   The platelet pellets were suspended and washed five times with 0.1 M phosphate buffer.   The fixed platelets, resuspended in 0.01 M phosphate buffered saline (PBS), were incubated with rabbit anti-GPIIb/IIIa IgG (1:10 final dilution)[7] for 60 min at room temperature.   After centrifugation at 3,000 g for 2 min, the platelets were washed with PBS three times as described above.   The platelet suspensions in PBS then were incubated with protein A coupled to 5 nm colloidal gold (protein A-gold, 1:10 final dilution) for 60 min at room temperature.   After the centrifugation, the platelet pellets were post-fixed with 1% osmium tetroxide and were prepared for electron microscopy according to conventional method as described above.   For control experiments, nonimmune normal rabbit IgG and protein A-gold were used.

To determine the localization of actin and myosin after the simultaneous extraction and fixation of the DTSP-pretreated platelets, the pellets were dehydrated with a graded ethanol series and finally embedded in Lowicryl K4M at -20°C as reported previously.[8,9]   Thin sections of Lowicryl K4M-embedded samples were prepared and mounted on nickel grids.   After blocking with 0.5% BSA, the sections were incubated with a rabbit anti-actin antibody (1:10 dilution) or a rabbit anti-myosin antibody (1:50 dilution) for 16 hr at room temperature, washed with PBS, and incubated with a goat anti-rabbit IgG coupled to 10 nm colloidal gold (1:20 dilution) for 3 hr at room temperature.   For control experiments, nonimmune normal rabbit IgG and the immunogold were used.   After washing, all sections were stained with uranyl acetate and lead citrate and then examined in a electron microscope.

RESULTS

Unstimulated platelets   After simultaneous extraction and fixation following DTSP-pretreatment, unstimulated platelets appeared discoid-shaped.   They contained some of the plasma membrane and abundant cytoskeletal structures.   Many microfilaments were dispersed at random in the cytoplasm.   Microtubules formed a circumferential band and were located at the poles of platelet.   Approximately 20-40% of the whole plasma membrane was resistant to the detergent and remained.   The residual membrane appeared as an intact unit membrane with a trilaminar structure (Fig. 1).   Amorphous structures were observed with 10-70 nm in diameters distributed at 20-100 nm intervals on the surface of plasma membrane.   All these amorphous structures appeared to penetrate to the membrane and to be connected with the submembrane zone just beneath the plasma membrane, from which microfilaments extended to the cytoplasm.   Similar amorphous structures were observed on the surface membrane of the unextracted platelets prepared for immunocytochemical studies of GPIIb/IIIa (Fig. 2).   Gold particles which show the localization of GPIIb/IIIa were present on the amorphous structures and were 10 to 20 nm apart from the outer surface of the plasma

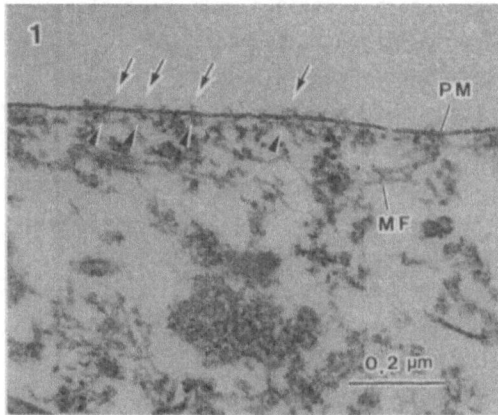
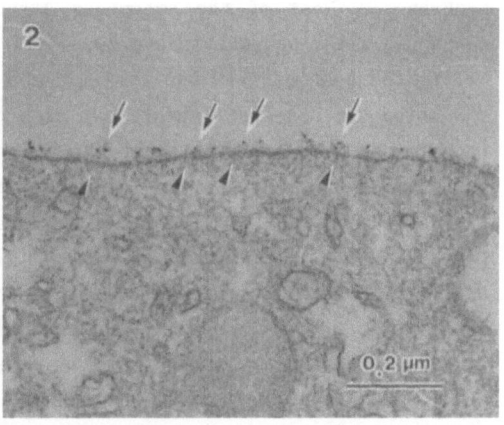

Fig. 1. Cytoskeletons of a resting platelet. A part of the plasma membrane of an unstimulated platelet which was crosslinked with DTSP prior to simultaneous extraction and fixation. The plasma membrane (PM) is observed as an unit membrane with a trilaminar structure. Amorphous structures are present on the plasma membrane (arrows). They penetrate plasma membrane and are connected with the membrane skeleton just beneath plasma membrane (arrowheads). MF, microfilaments.

Fig. 2. Localization of GPIIb/IIIa in an unstimulated platelet. Intact platelets were incubated with a rabbit anti-GPIIb/IIIa antibody followed by protein A-gold. Gold particles are distributed in the amorphous structures on the plasma membrane (arrows). Most of the amorphous structures that contain gold particles are connected with the membrane skeleton (arrowheads).

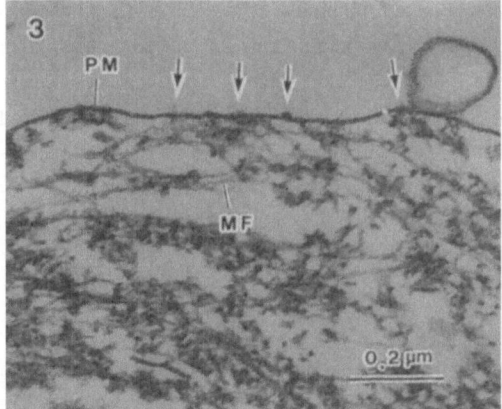
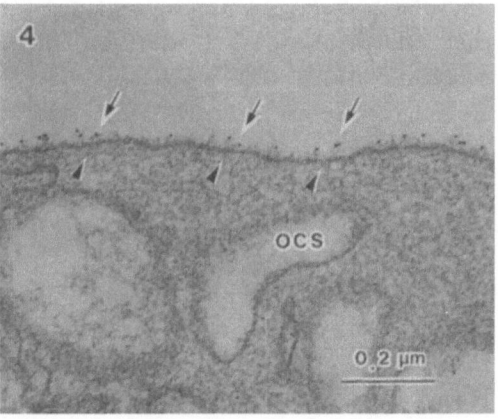

Fig. 3. Cytoskeletons of an activated platelet. After activation with thrombin for 15 sec, the platelets were crosslinked with DTSP prior to simultaneous extraction and fixation. Amorphous structures on the plasma membrane (PM) are connected with the membrane skeleton just beneath the plasma membrane (arrows). Long microfilaments extending from the membrane skeleton increase and form dense networks around granules. MF, microfilaments.

Fig. 4. Localization of GPIIb/IIIa in an activated platelet. Thrombin-activated platelets were incubated with a rabbit anti-GPIIb/IIIa antibody followed by a protein A-gold. Gold particles are distributed in the amorphous structures on the plasma membrane (arrows). Some of the amorphous structures that contain gold particles are connected with the membrane skeleton (arrowheads). OCS, open canalicular system.

membrane.   Most of the amorphous structures which contained gold particles were connected with the submembrane zone just beneath the plasma membrane.

In the sections stained for actin, gold particles were distributed not only on the cytoplasmic microfilaments but also on the submembrane zone at the periphery of the discoid platelets (data not shown).   Therefore, at least some of the microfilaments in the cytoplasm were identified as actin filaments.   Moreover, the result indicates that the submembrane zone observed in the sample of extracted platelets corresponds to the membrane skeleton because the zone contained actin.[3]   Thus, we hereinafter refer to this submembrane zone as the membrane skeleton.   On the other hand, very few gold particles that showed the localization of myosin were detected on the microfilaments or the membrane skeleton (data not shown).

Activated platelets   The thrombin-activated platelets were round in shape. The amorphous structures which were always associated with the membrane skeleton were observed on the surface of the plasma membrane (Fig. 3).   Microfilaments extending from the membrane skeleton increased and formed dense networks.   Triton-insoluble granules were surrounded by the dense networks of microfilaments in the central part of the platelets.

In the samples probed with antibody against GPIIb/IIIa, gold particles were abundantly present on the surface of the activated platelets (Fig. 4).   Although the number of gold particles somewhat increased on the surface,   the patterns of distribution of gold particles did not show marked changes compared with those in unstimulated platelets.   Most of the gold particles were located on the amorphous structures which were connected with dense membrane skeleton just beneath plasma membrane.   In the sections stained for actin, many gold particles were distributed on the microfilaments forming the dense networks surrounding the centralized granules (data not shown).   Myosin were localized on the microfilaments as similar as the localization of actin (data not shown).

DISCUSSION

In the present study, we have demonstrated that a part of the amorphous structures on the plasma membrane are GPIIb/IIIa molecules.   With careful inspection, we found that the amorphous structures containing GPIIb/IIIa molecule penetrated to the plasma membrane and connected with the submembrane zone.   The submembrane zone was identified as membrane skeleton because actin was detected in the zone.   Thus, these results indicate an association of GPIIb/IIIa with the membrane skeleton morphologically.   Microfilaments were observed abundantly in the cytoplasm of both unstimulated and thrombin-activated platelets.   Furthermore, the 20-40% of whole plasma membrane preserved almost intact and amorphous structures on the plasma membrane could be observed after the crosslinking prior to detergent extraction using Triton X-100.   The amorphous structures appeared to be similar to the glycocalyx which corresponds to glycoproteins.

It is well known that actin polymerization increases after platelet acti- vation.[10,11] In the present study, microfilaments extending from the membrane skeleton increased and formed dense networks around Triton-insoluble granules after activation. Moreover, actin was detected in the microfilaments. Our results indicate an increase in actin filaments extending from the membrane skeleton of platelets during activation. We found that myosin became associated with actin filaments after the activation, suggesting the formation of actomyosin. This find- ing was in agreement with the reports by Tanaka ct al.[12] and Nakata and Hirokawa.[13]

In both the unstimulated and thrombin-activated platelets, we found that amor- phous structures 10-70 nm in diameter were distributed at 20-100 nm intervals on the surface of the plasma membrane in the simultaneously extracted and fixed samples. Moreover, similar amorphous structures were observed on the plasma membrane, most of which contained gold particles, in the intact platelets prepared for immunocytochem- istry of GPIIb/IIIa. These results suggest that a part of the amorphous structures correspond to the GPIIb/IIIa complex. Although previous investigator has reported the association between extracellular coat and the membrane skeleton,[14] the GPIIb/IIIa was not identified in the coat. The GPIIb/IIIa complex is the most abundant platelet cell-surface protein. A normal platelet contains approximately 50,000 GPIIb/IIIa complexes, which present about 1 to 2% of the total platelet protein. This complex acts as a binding site for fibrinogen on activated plate- lets, and is essential for platelet aggregation. Previous investigators have found that the GPIIb/IIIa formed clusters on the plasma membrane in the activated plate- lets using isolated membranes[15] or surface replica technique.[16] In the present study, however, we could not observed clusters composed by the GPIIb/IIIa molecule except increase of the molecule in the thrombin-activated platelets.

A biochemical study have reported that a small percentage of the GPIIb/IIIa appears to be associated with actin filaments in unstimulated platelets using bio- chemical methods.[17] The association between GPIIb/IIIa and actin may be mediated by several cytoskeletal proteins. There is no morphological evidence for the asso- ciation of the GPIIb/IIIa with the membrane skeleton. We found that amorphous structures on the plasma membrane, which were identified as the GPIIb/IIIa complex by an immunocytochemical method, penetrated to plasma membrane and connected with membrane skeleton. This result demonstrates morphologically that the GPIIb/IIIa has a transmembrane portion and is associated with the membrane skeleton.[18] Actin was immunocytochemically detected in the membrane skeleton, suggesting that the GPIIb/IIIa was connected with actin in the membrane skeleton. We found that micro- filaments in cytoplasm were connected with the membrane skeleton and these associa- tions increased during platelet activation. The subsequent interaction of the membrane skeleton and the actin filaments with the GPIIb/IIIa complex may play an essential role in the process of platelet activation.

REFERENCES

1. Harris H. (1981) Regulation of motile activity in platelets. Platelets in Biology and Pathology Vol. 2. Elsevier North-Holland Biomedical Press, Amsterdam. pp 474-500
2. Fox J.E.B. and Phillips D.R. (1983) Polymerization and organization of actin filaments within platelets. Semin. Hematol. 20: 243-259
3. Fox J.E.B., Boyles J.K., Berndt M.C., Steffen P.K. and Anderson L.K. (1988) Identification of a membrane skeleton in platelets. J. Cell. Biol. 106: 1525-1538
4. Suzuki H., Yamamoto N., Tanoue K. and Yamazaki H. (1987) Glycoprotein Ib distribution on the surface of platelets in resting and activation states: an electron microscope study. Histochem. J. 19: 125-136
5. Bell P.B. (1981) The application of scanning electron microscopy to the study of the cytoskeleton of cells in culture. Scanning Electron Microscopy Vol II. SEM Inc, Chicago, pp 139-157
6. Boyles J.K., Fox J.E.B., Phillips D.R. and Stenberg P.E. (1985) Organization of the cytoskeleton in resting, discoid platelets: preservation of actin filaments by a modified fixation that prevents osmium damage. J. Cell. Biol. 101: 1463-1472
7. Tanoue K., Hasegawa S., Yamaguchi A., Yamamoto N. and Yamazaki H. (1987) A new variant of thrombasthenia with abnormally glycosylated GPIIb/IIIa. Thromb. Res. 47: 323-333
8. Suzuki H., Kinlough-Rathbone R.L., Packham M.A., Tanoue K., Yamazaki H. and Mustard J.F. (1988) Immunocytochemical localization of fibrinogen on washed human platelets. Lack of requirement for fibrinogen during adenosine diphosphate-induced responses and enhanced fibrinogen binding in a medium with low calcium levels. Blood 71: 850-860
9. Suzuki H., Katagiri Y., Tsukita S., Tanoue K. and Yamazaki H. (1990) Localization of adhesive proteins in two newly subdivided zones in electron-lucent matrix of human platelet α-granules. Histochem. 94: 337-344
10. Carlsson L., Markey F., Blikstad I., Persson T. and Lindberg U. (1979) Reorganization of actin in platelets stimulated by thrombin as measured by the DNase I inhibition assay. Proc. Natl. Acad. Sci. 76: 6376-6380
11. Jennings L.K., Fox J.E.B., Edwards H.H. and Phillips D.R. (1981) Changes in the cytoskeletal structure of human platelets following thrombin activation. J. Biol. Chem. 256: 6927-6932
12. Tanaka K., Shibata N., Okamoto K., Matsusaka T., Fukuda H., Takagi M., Fujii N., Toya N. and Onji T. (1986) Reorganization of myosin in surface-activated spreading platelets. J. Ultastruct. Mol. Struct. Res. 97: 165-186
13. Nakata T. and Hirokawa N. (1987) Cytoskeletal reorganization of human platelets after stimulation revealed by the quick-freeze deep-etch technique. J. Cell. Biol. 105: 1771-1780
14. Bearer E.L. (1990) Platelet membrane skeleton revealed by quick-freeze deep-etch. Anat. Rec. 227: 1-11
15. Polley M.J., Leung L.L.K., Clark F.Y. and Nachman R.L. (1981) Thrombin-induced platelet membrane glycoprotein IIb and IIIa complex formation. An electron microscope study. J. Exp. Med. 154: 1058-1068
16. Isenberg W.M., McEver R.P., Phillips D.R., Shuman M.A. and Bainton D.F. (1987) The platelet fibrinogen receptor: an immunogold-surface replica study of agonist-induced ligand binding and receptor clustering. J. Cell. Biol. 104: 1655-1663
17. Painter R.G., Prodouz K.N. and Gaarde W. (1985) Isolation of a subpopulation of glycoprotein IIb-III from platelet membranes that is bound to membrane actin. J. Cell. Biol. 100: 652-657
18. Suzuki H., Tanoue K. and Yamazaki H. (1991) Morphological evidence for the association of plasma membrane glycoprotein IIb/IIIa with the membrane skeleton in human platelets. Histochem. 96: 31-39

MOLECULAR MECHANISM OF SHEAR STRESS-INDUCED PLATELET AGGREGATION

Y.IKEDA, M.HANDA, Y.KAWAI AND K. WATANABE

Department of Hematology and Laboratory Medicine, School of Medicine, Keio University, 35 Shinanomachi Shinjuku-ku Tokyo Japan, 160

INTRODUCTION

Blood circulating in vessels is exposed to shear stress caused by the force necessary to produce flow; the difference in velocity between layers situated at varying distances from the vessel wall determines the shear rate, which is directly proportional to the shear force and inversely proportional to the viscosity of blood. Previous studies have demonstrated that exposure of platelets to shear stress leads to aggregation in the absence of exogenous agonists(1)(2). Under high shear stress, which may mimic the rheological situation existing in certain districts of the arterial circulation, von Willebrand factor (vWf) may have a crucial role in the formation of the irreversible platelet aggregates(3)(4). However, the precise molecular mechanism of shear stress-induced platelet aggregation (SIPA) has yet to be clarified.

Recently we have developed a mechanical device to measure SIPA, using a cone and plate type viscometer adapted for recording of both light transmission and fluorescence intensity. The machine has several advantages to measure SIPA; 1. easy to handle with a small volume of sample, 2. continuous recording is possible with good reproducibility, 3. simultaneous measurement of intracellular calcium ion, $[Ca++]i$, during SIPA is also possible. Using this apparatus, we have obtained interesting results as described in this paper.

Low shear forces induce formation of reversible platelet aggregates dependent on fibrinogen interaction with GPIIb/IIIa, while more stable aggregates are formed at high shear forces, which is independent of fibrinogen, but requires vWf binding to both GPIb/IX and GPIIb/IIIa. No other adhesive proteins such as fibrinogen seems to be involved in aggregation induced by high shear stress. It was also found that binding of vWf to GPIb induced by shear stress causes influx of calcium ions, which appears prerequisite for aggregation.

METHODS

Patients:
The patients with thrombasthenia, Bernard-Soulier syndrome, afibrinogenemia and severe von Willebrand disease fulfilled all the accepted criteria for the diagnosis of these disorders. Thrombasthenic patients had no measurable GPIIb/IIIa complex. Patients with Bernard-Soulier syndrome had no measurable GPIb. The patients with afibrinogenemia and severe von Willebrand disease had less than 10 ug/ml of plasma fibrinogen and less than 0.1% of normal plasma vWf level.

Modified Viscometer for Measuring SIPA and [Ca++]i :
   The cone and plate type viscometer used in this study is
similar to the one described previously (5). The entire system
is schematically shown in Fig. 1.

An Apparatus

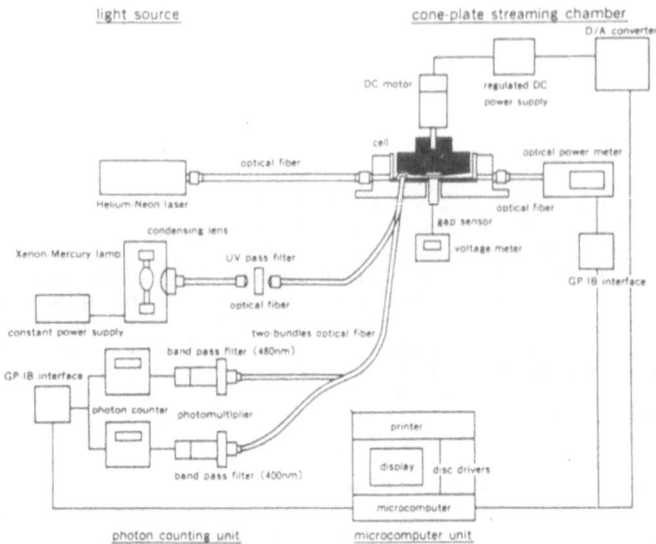

Fig.1. Schematic drawing of the apparatus to measure shear-
induced platelet aggregation and intracellular calcium
concentration simultaneously.

   For measuring shear-induced aggregation, 400 ul of platelet-
rich plasma was applied onto the surface of the plate and was
exposed to shear stress regulated by the rotation speed.
Aggregation was monitored continuously by recording the intensity
of the light transmitted through the platelet suspension. For
measuring [Ca++]i, indo-1AM loaded washed platelets were applied
to the center of the plate in the presence of 10 ug/ml of vWf,
100 ug/ml of fibrinogen and 1 mM CaCl2 and exposed to varying
shear stress. Fluorescence light intensity was then monitored
at 400 nm and 480 nm. The fluorescence received by the
photomultiplier tube was then calculated by the photon counting
method. Change of [Ca++]i expressed as uM was shown on the
display at the microcomputer unit.

Monoclonal antibodies:
   All the monoclonal antibodies used in this study were obtained
and characterized as described previously except for nmc-4 (6).
Nmc-4 is a murine IgG obtained by immunization with purified vWf.
The antibody, which reacts with 52/48 kDa fragment of vWf
containing the GPIb binding domain, inhibits ristocetin-induced
platelet aggregation completely.

RESULTS

Aggregation was absent in two patients with thrombasthenia both
at low (12 dyn/cm2) and high shear (108 dyn/cm2); in contrast,
aggregation was normal at low but absent at high shear in two

patients with Bernard-Soulier syndrome. In one patient with
afibrinogenemia, aggregation was absent at low but normal at high
shear. Two patients with a complete deficiency of vWf revealed
normal aggregation at low but defective aggregation at high
shear.
(Fig. 2.)

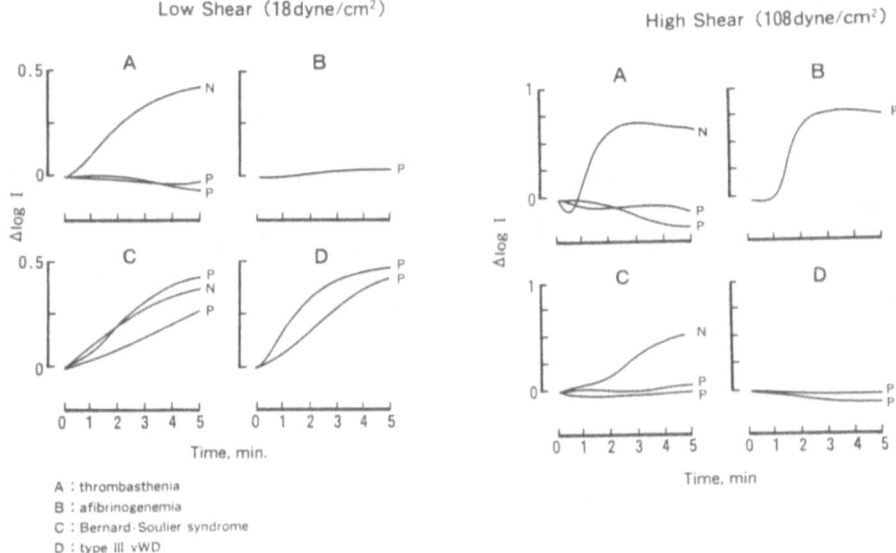

A : thrombasthenia
B : afibrinogenemia
C : Bernard-Soulier syndrome
D : type III vWD

Fig.2   Shear-induced platelet aggregation in patients with
congenital deficiencies of membrane glycoproteins and plasma
adhesive proteins.

To further explore the effects of shear forces on platelet
aggregation, a gradient varying between 6 and 108 dyn/cm2 was
applied to citrated platelet-rich plasma over a 5-min period.
Two distinct peaks were obtained as shown in Fig.3. While shear
stress varied between 6 and 12 dyn/cm2 in 90 sec, light
transmittance increased, indicating the formation of platelet
aggregates. In the ensuing 120 sec, the gradient varied between
12 and 108 dyn/cm2. This resulted in a decrease in light
transmittance followed by prominent second increase, which
started approximately when the shear stress reached 81 dyn/cm2 at
176 sec. To confirm the distinct roles of different platelet
membrane glycoproteins and plasma adhesive proteins in
this type of aggregation, the effects of three different
monoclonal antibodies were tested. Both peaks were completely
abolished by LJ-CP8, which reacts against GPIIb/IIIa complex to
inhibit agonist-induced platelet aggregation. The anti-GPIb
antibody LJ-Ib1, which inhibits vWf binding to GPIb, blocks only
second large peaks. Similar results were obtained with anti-vWf
monoclonal antibody LJ-152B/6, which selectively inhibits binding
of vWf to GPIIb/IIIa.
Aggregation and [Ca++]i was simultaneosly measured using indo-
1AM loaded washed platelet suspensions in the presence of vWf,
fibrinogen and CaCl2 as described in the Method.
No change of [Ca++]i was observed during low shear-induced
platelet aggregation, while gradual rise in [Ca++]i was seen
concurrently with or shortly before vWf-dependent aggregation
occurred. (Fig.4)

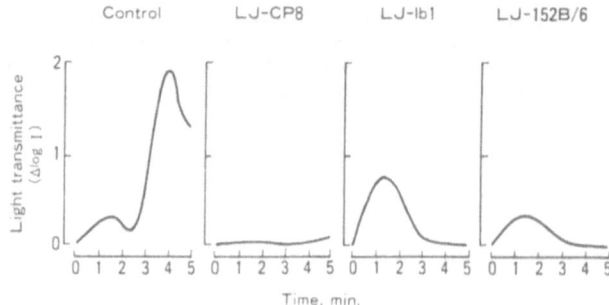

Fig.3 Effects of monoclonal antibodies on aggregation induced by varying shear stress.

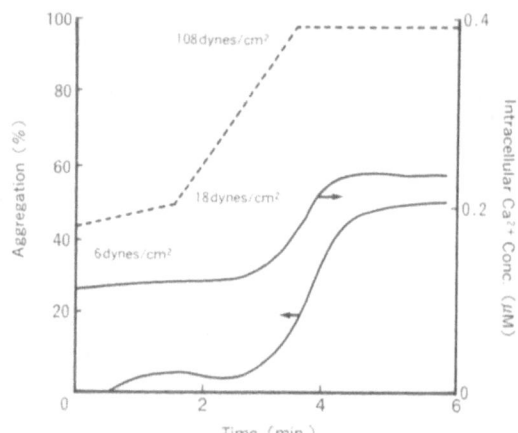

Fig.4 Simultaneous measurement of aggregation and [Ca++]i under varying shear stress.

No change of [Ca++]i was observed when EGTA (1mM) was present in the medium instead of CaCl2 indicating that a rise in [Ca++]i was due to influx, not due to mobilization inside platelets. To confirm that the rise in [Ca++]i was not a consequence of aggregation, thrombasthenic platelets were used in the same experimental conditions. While aggregation was totally abolished, a rise in [Ca++]i was even more pronounced than that in normal platelets during the late phase of aggregation.
To elucidate the mechanism of Ca++ influx, effects of monoclonal antibodies on both aggregation and calcium influx were tested.
Both LJ-152B/6 and nmc-4 abolished high shear-induced vWf-mediated aggregation, but calcium influx was inhibited only by nmc-4.
LJ-Ib1 also inhibited both aggregation and calcium influx.

DISCUSSION

Our results characterize the distinct adhesive proteins and membrane receptors involved in platelet aggregation occurring under varying shear stress. The molecular mechanisms mediating platelet-to-platelet contact are different; fibrinogen

interaction with GPIIb/IIIa appears necessary at low shear, while
vWf interaction with both GPIb/IX and GPIIb/IIIa, but independent
of fibrinogen, is required at high shear. (Fig.5)

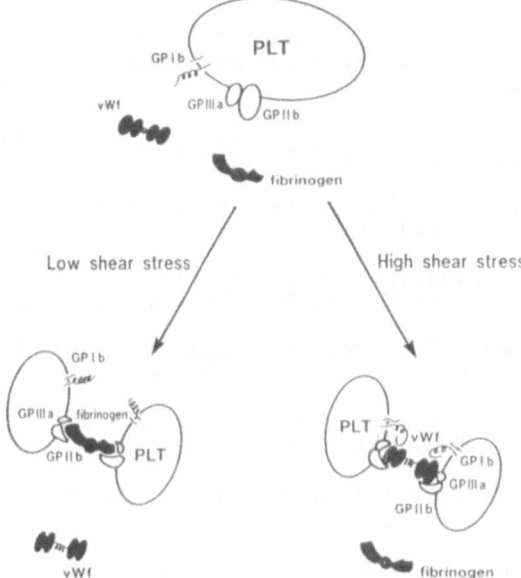

Fig.5   Possible mechanism of shear-induced platelet aggregation.

When platelet-rich plasma contains normal levels of divalent
cations, aggregation is negligible at low shear values but
conspicuous above 81 dyn/cm2, even though its maximal extent is
one-fourth to one-third of that seen in citrated plasma (data
not shown).   To define the intracellular events responsible for
induction of binding of adhesive proteins to their respective
membrane receptors, we have also measured [Ca++]i during SIPA,
using a cone-plate type viscometer.   Under low shear, no change
of [Ca++]i was observed.   In contrast, binding of vWf to GPIb
induced by high shear caused the influx of Ca++ ions.   It
appears that Ca++ ions influx is essential for the induction of
aggregation.
   Under high shear stress conditions, attachment of platelets to
surface-bound vWf through GPIb/IX is also crucial for initial
adhesion (7).   Our results suggest that the interaction of vWf
with GPIb/IX may trigger a platelet response leading to
expression of the binding function of GPIIb/IIIa which, by
binding vWf from surrounding medium, can strengthen the
interplatelet contact to sustain the opposing effect of shear
forces.   The interplatelet interaction leading to thrombus
growth is initially GPIb-dependent, but becomes progressively
reinforced by expression of the binding function of GPIIb/IIIa.
It is fascinating to speculate that this self-sustaining process
involving two membrane glycoproteins and one adhesive ligand is
operating during thrombogenesis in vivo.
   In acute arterial occlusion, platelets in circulating blood may
become exposed to shear stress values in excess of several
hundred dyn/cm2, as in small arteries and arterioles partially
obstructed by atherosclerotic processes or as a consequence of
vasospasms (8)(9).   Strategies directed at inhibiting vWf
interaction with platelets may be useful in preventing thrombotic
vascular occlusion.

REFERENCES

1.	Hung,T.C., Hochmuth,R.M., Joist,J.H. and Stera,S.P.ʼ (1976)
Trans. Am. Soc. Artif. Intern. Organs 22: 285-291.

2.	Ikeda,Y., Murata,M., Araki,Y., Watanabe,K., Ando,Y.,
Itagaki,I., Mori,Y., Ichitani,M. and Sakai,K. (1988)
Thromb. Res. 51: 157-163.

3.	Peterson,D.M., Stathopoulos,N.A., Giorgio,T.D.,
Hellums,J.D. and Moake,J.L. (1987) Blood 69: 625-628.

4.	Moake,J.L., Turner,N.A., Stathopoulos,N.A., Nolasco,L.,
Hellums,J.D. (1988) Blood 71: 1366-1374.

5.	Fukuyama,M., Sakai,K., Murata,M., Kawai,Y. Watanabe,K.
Handa,M. and Ikeda,Y. (1989) Thromb. Res. 54: 253-260.

6.	Ikeda,Y., Handa,M., Kawai,Y., Watanabe,K. and Ruggeri,Z.M.
(1991) J. Clin. Invest. 87: 1234-1240.

7.	Weiss,H.J., Turitto,V.T. and Baumgartner,H.R. (1978)
J. Lab. Clin. Med. 92: 750-764.

8.	Back,C.H., Radbill,J.R. and Crawford. (1977) J. Biomech.
10: 339-353.

9.	Lipowski,H.H., Usami,S. and Chien,S. (1980) Microvasc.
Res. 19: 297-319.

THE EFFECTS OF LIPID PEROXIDE ON BLOOD PLATELET SEROTONIN

T.HIRAMITSU[1]   AND M.H.PIETRASZEK[2]

[1]Department of Ophthalmology and [2]Physiology, Hamamatsu University School of Medicine, Shizuokaken, Hamamatsu-shi, 3600 Handa-cho, 431-31 Japan

INTRODUCTION

The pathophysiological roles of serotonin (5-hydroxytryptamine; 5HT) in cardiovascular diseases as well as in psychiatric syndromes such as, for instance, depression, have been documented. Serotonin is released from aggregating platelets. Platelet aggregation may play a role in the process of myocardial ischemia and/or infarction.[1]

Lipid peroxides (LPO) give the deleterious effects on cells and tissues and cause many diseases. Since one of the deleterious effects may be ascribed to platelet activation induced by LPO, the relationship between lipid peroxides and vascular diseases involved in myocardial infarction and/or cerebral stroke has gained attention in clinical field. To clarify the relationship, several studies have been done. In in vitro experiment, a rapid and marked platelet aggregation response was caused by peroxized arachidonic acid whereas a slow response resulted from exposure to free fatty acid.[2] The contraction of the isolated artery was markely induced by 15-hydroperoxy arachidonic acid.[3] When linoleic acid peroxide (13-hydroperoxy linoleic acid) was injected inside the ventricle of the heart, experimental myocardial infarction was demonstrated by electrocardiograph and scanning electron microscope method. But no change was observed by the injection of linoleic acid.[4] Futher,Yagi et al showed that drastic morphological changes in the intima and endothelial cells of the aorta were provoked by a single injection of linoleic acid hydroperoxide but not by the injection of linoleic acid.[5]
This paper describes the effects of LPO on serotonin release in the rabbit blood in vivo.

THE CHANGES IN SEROTONIN LEVEL IN WHOLE BLOOD AND IN PLASMA BY LPO INJECTION

13-Hydroperoxy linoleic acid (LPO) was refined from linoleic acid incubated with soybean lipoxygenase according to the method of Egmond et at.[6] LPO resolved in borate buffer (1mg/kg) was injected intravenously into adult albino rabbits. As controls, linoleic acid solution(1mg/kg) was injected. The blood samples were taken before the injection and at 1 , 3 and 5 hr. after the i.v. injecttion. For the measurement of setotonin release from platelets, a suspension of platelets was incubated with 10 M of serotonin for 15 min. The quantitative analysis of serotonin in whole blood and plasma was performed by HPLC with a flurorescence detector according to Anderson et al.[7]

Figures 1 and 2 illustrate the effect of linoleic acid on serotonin concentration in whole blood and plasma, respectively. No significant changes were induced by linoleic acid.

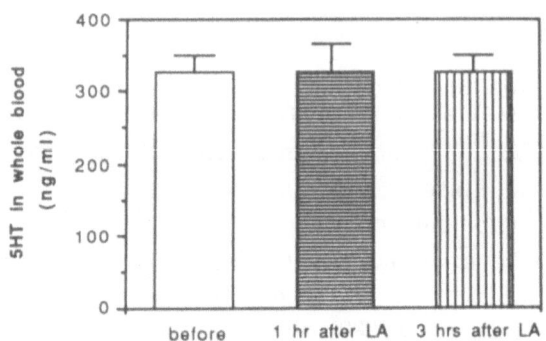

Fig.1. The effect of linoleic acid (LA) on 5HT concentration in whole blood.

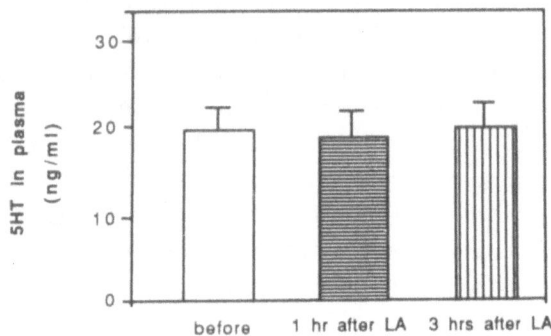

Fig.2. The effect of linoleic acid (LA) on 5HT concentration in plasma.

Figures 3 and 4 illustrate the changes of serotonin concentration by LPO. Serotonin concentration was reduced by 60% in whole blood and was increased two folds in plasma at 1 hr after LPO injection. The changes almost returned to control at 5 hr after LPO injection. These findings may be ascribed to serotonin released from platelet markedly activated by LPO.

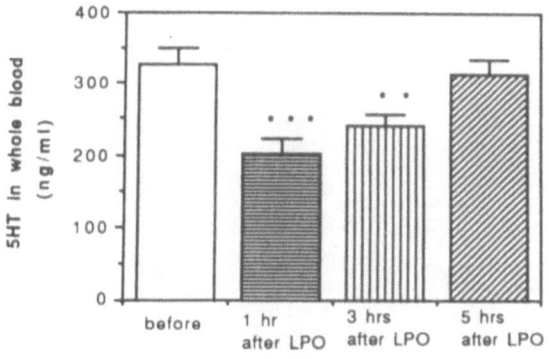

Fig.3. The effect of LPO on 5HT concentration in whole blood.
*** P<0.001, ** P<0.01 in respect control (before).

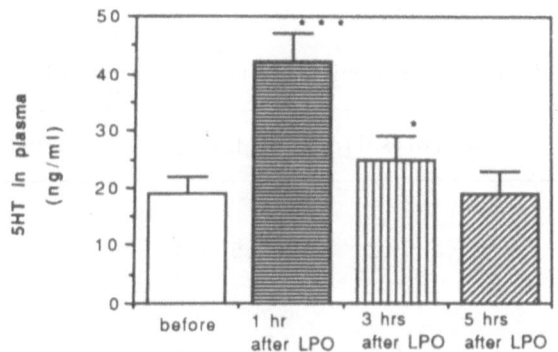

Fig.4. The effect of LPO on 5HT concentration in plasma.
  *** p<0.001, *p<0.01 in respect to control (before).

THE EFFECTS OF ASPIRIN ON LIPID PEROXIDE INDUCED CHANGES IN SEROTONIN CONCENTRATION.

Aspirin is a potent prostaglandin synthesis inhibitor by which the formation of thromoxan A_2 aggregating platelet can be inhibited. In order to study the role of platelet aggregation at the time of the changes of serotonin concentration induced by LPO, the effect of aspirin was examined. Aspirin DL-lysine resolved in distilled water (50mg/kg) was intravenously injected to the rabbit 1 hr before LPO injection. Figure 5 illustrates the effect of aspirin on LPO-induced changes in whole blood and plasma 5HT concentration. Aspirin significantly inhibited the change of 5HT concentration induced by LPO seen at 1 and 3 hr after the injection. These findings show that the prostaglandin-mediated platelet aggregation may be involved in LPO induced serotonin release.

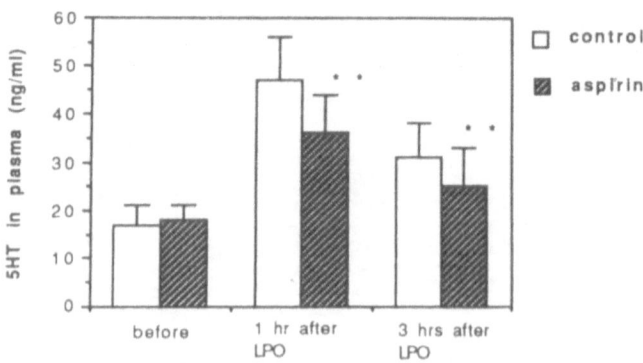

Fig.5. The effect of aspirin on LOP-induced changes in plasama 5HT concentration. ** p<0.01 in respect to control value (without aspirin).

THE EFFECTS OF PAF ANTAGONIST ON LIPID PEROXIDE INDUCED CHANGES IN SEROTONIN CONCENTRATION.

Table 1. The effect of LPO on serotonin release from blood platelets in vitro.

	control(saline)	LPO	significance
5 HT release (pmol/10^8 pl) after 30 min	14.1±3.2	14.5±3.8	N.S.
after 60 min of incubation	21.6±4.3	22.9±4.5	N.S.

values are mean±standard error.

As seen in Table 1, LPO when incubated with platelet suspension had no effect on serotonin release reaction. It could be suggested that increase of free serotonin in the plasma observed after LPO injection is not related to its effect on blood platelets. LPO may also stimulate white cells to release PAF which is considered as one of the most powerful platelet activating factor. Platelet from rabbits are so sensitive to PAF that they can be activated in concentrations as low as 1pM.[8] Thus, we used a newly synthesized PAF inhibitor, TCV-309(Takeda Pharmacol.Co.). It was i.v. injected 10 min. before LPO injection at dose of 1mg/kg. As expected, PAF antagonist completely inhibited LPO-induced increase in plasma serotonin (Figure 6).

Prostaglandins participating in platelet aggregation are produced by arachidonate peroxidation. Peroxidation has been considered to play a role in platelet aggregation. Several in vitro studies demonstrated the close

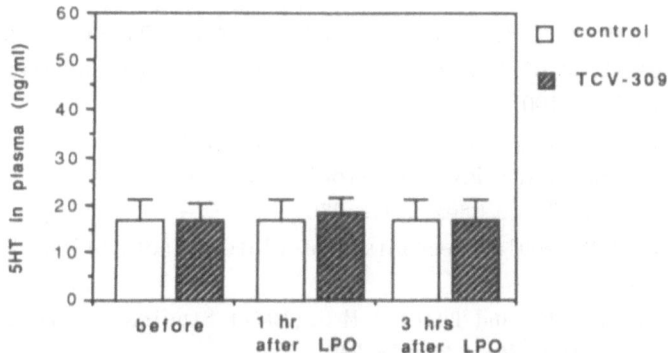

Fig.6. The effect of TCV-309 (PAF-antagonist) on LPO-induced changes in plasma 5HT concentration.

relationship between peroxidation and platelet activation. Lipid peroxides in human platelets were found when washed platelets were aggregated by thrombin.[9] Other studies demonstrated that α-tocopherol inhibited serotonin release from platelet and superoxide anion enhanced serotonin release from platelet.[10,11] These results show the direct relationship between peroxidation including oxygen radical and pletelet activation. Our studies showed the marked effect of LPO on the activation of platelets at alow dose (1mg/kg). We speculate that this effect may be due to the indirect action of LPO to accumulate white cells resulting in PAF release to which platelet is very sensitive. The important role of LPO in causing vascular diseases has been documented by many studies. LPO may initiate vascular diseases by activating platelet.

REFERENCES
1. J.I., Gershengorn K., Kranz P.D., and Oestreicher R.(1972) Protection against epinephrin-induced myocardial necrosis by drugs that inhibit platelet aggregation. Am.J.Cardiol. 30:838-843.
2. Mickel H.S. and Horbar J.(1974) The effect of peroxidized arachidonic acid upon human platelet aggregation. Lipids. 9:68-71.
3. Koide T.,Neichi T., Takato M., Matsushita H., Sugioka K., Nakano M., and Hata S.(1982) Possible mechanism of 15-hydroperoxyarachidonic acid-induced contraction of the canine basilar artery in vitro. J. Pharmacol. Exp. Ther.2 21:481-488.
4. Mizukami M., Ano J., Sakai K., Hata S.and Nakano M.(1984) Experimental myocardial infarction:effects of a lipid peroxide, 13-hydroperoxy linoleic acid on coronary circulation in rats. Arzneim-Forsh./Drug Res.34:569-571.
5. Yagi K., Ohkawa H., Ohishi N., Yamsaki M. and Katou M. (1981) Lesion of aortic intima caused by intravenous administration of linoleic acid hydroperoxide. J. Appl. Biochem. 3:58-65.

6. Egmont M.R., Brunori M., and Fasella P.M.(1976) The steady-state kinetics of the oxygenation of linoleic acid catalysed by soybean lipoxygenase. Eur. J. Biochem. 61:93-100.
7. Anderson G.M., Young J.G., Ohen D.J., Schlicht K.R. and Patel N.(1981) Liquid-chromatographic determination of serotonin and tryptophan in whole blood and plasma. Clin. Chem 5:775-776.
8. Siess W.(1989) Molecular mechanisms of platelet activation. Physiological. Rev. 69:58-177.
9. Okuma M., Steiner M. and Baldini M.G.(1971) Studies on lipid peroxides in platelets. J.Lab.Clin.Med. 77:728-742.
10. Steiner M. and Anastasi J.(1976) Vitamin E. An inhibitor of the platelet release reaction. J.Clin.Invest. 57:732-737.
11. Handin R.I., Karabin R. and Boxer G.J.(1977) Enhancement of platelet function by superoxide anion. J.Clin.Invest.59:959-965.

LIPOPROTEIN (A) (LP(A)): ITS RELEVANCE TO DIABETIC ANGIOPATHY

T. TAMINATO, F-N. HU, T. TOMINAGA, S. SEN, T. YOSHIMI, Y. SUZUKI, M. NAGAO, Y. TAKADA AND A. TAKADA

The Second Department of Medicine, Department of Physiology, Hamamatsu University School of Medicine and Haibara General Hospital, SHIZUOKA, 431-31, JAPAN

INTRODUCTION

Lipoprotein(a)(Lp(a)) was first discovered by Berg as a low-density lipoprotein like molecule in 1963. Much of the interest in the Lp(a) stems from several clinical studies that showed a highly significant association between elevated levels of the lipoprotein and susceptibilitiy to coronary artery disease(1). Compelling evidences accumulated in these 10 years indicate that Lp(a) is a major risk factor for atherosclerosis, coronary heart disease and cerebrovascular disease. Lp(a) is now believed to be independent from serum cholesterol and other risks. Lp(a) is composed of a lipid core, low-density lipoprotein containing an apoprotein B-100 subunit, and a unique lipoprotein "apoprotein(a)(apo(a))". The apo(a) subunit is linked by a disulfide bond to the apo B subunit.

A striking structural similarity of the serine protease and kringle-5 in apo(a) to plasminogen, arouse a great interest concerning the pathogenic role of Lp(a) in atherosclerosis and thrombogenesis(2). It contains a serine protease domain, one copy of the kringle-5 region, and 37 copies of the kringle-4 domain. In general, the kringle domains serve as regulatory domains and are important for the activation of specific proteases such as plasminogen and tissue plasminogen activator(t-PA). In addition, Lp(a) like immunoreactivity has been identified in the endothelium and intima of atherosclerotic coronary arteries(3). Therefore, it is possible that Lp(a) interfere with the fibrinolytic system because of its structural homology to plasminogen.

Diabetes mellitus is a syndrome characterized by hyperglycemia and its chronic complications, namely retinopathy and nephropathy, as results of microangiopathy. Although the pathogenesis of diabetic angiopathies remained unclear, exaggerated blood coagulation and blunted fibrinolysis have been observed in diabetes mellitus.

The present study was carried out to investigate the role of Lp(a) in the development of diabetic micro- and macroangiopathies.

METHODS

Plasma samples were taken from two groups of patients. In the first series of experiment, plasma samples were drawn from 60 patients tested for coronary angiography (16 diabetics and 44 non-diabetics). After the angiography, they were classified by the number of arteries occluded, and by the presence or absence of diabetes. In the second series of the study, 52 diabetics (28 male and

24female) with or without microangiopathy (retinopathy or nephropathy) were subjected to Lp(a) measurement. Diabetic retinopathy was diagnosed by ophthalmologists. Diabetic nephropathy was judged by continuous albuminuria. Albuminuria more than 300 mg/g.Cr was used as a criteria of overt nephropathy, while microalbuminuria (ranging from 20 to 300 mg/g.Cr) as a early stag of the nephropathy. Plasma Lp(a) was measured by a commercial kit ("TintElize Lp(a)", biopool, Sweden). The concentration of plasma tissue plasminogen activator (t-PA) was determined by an enzyme immunoassay (EIA) described elsewhere(4).

Statistical analysis was performed by ANOVA and Student's t test.

RESULTS

Plasma Lp(a) levels of diabetics were higher than that of non-diabetics (400.7 ± 71.4 vs. 317.4 ± 35.4 mg/l, Means $\pm$ SEM, p<0.05). In non-diabetics, patients with coronary artery occlusion showed higher Lp(a) levels than those without occlusion. Lp(a) levels in the patients with coronary vessel-involvement showed a stepwise elevation as the number of occlusion increased (non-occlusion group; 231.7 ± 48.5, 1-vessel occlusion; 335.4 ± 52.3, 2-vessel occlusion; 579.3 ± 97.2, 3-vessel occlusion 456.7 ± 193.1 mg/l,) (Fig.1, right panel). In diabetics, likewise non-diabetics, patients with coronary artery occlusion had significantly higher Lp(a) levels than non-affected group (Fig.1, left panel).

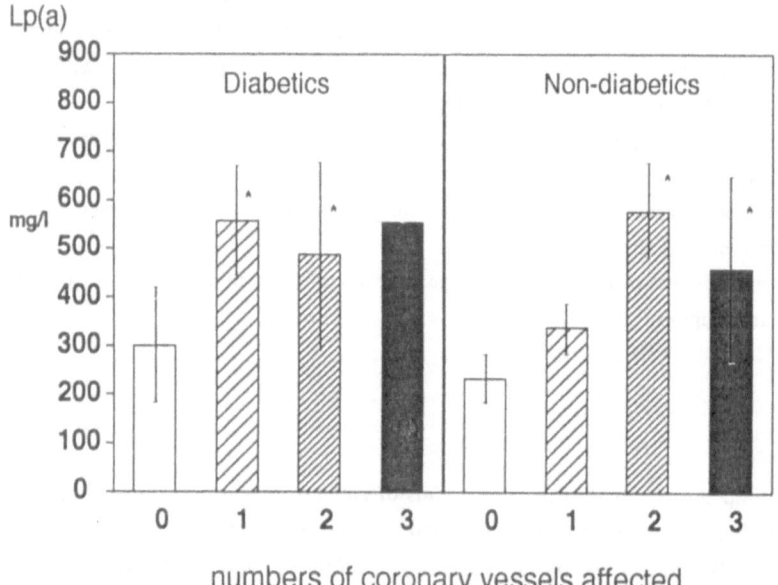

Fig.1 Plasma Lp(a) levels in patients tested for coronary angiography.　　　(* P< 0.05)

Plasma concentration of diabetics with retinopathy was significantly higher than that of the patients without retinopathy (Fig. 2). Diabetics complicated with diabetic nephropathy exhibited increased Lp(a) levels compared to non-proteinuric diabetics (Fig. 3). Especially, patients with overt nephropathy had a significantly higher Lp(a) concentration than non-proteinuric patients, while diabetics with microalbuminuria showed a slight elevation of the lipoprotein.

The correlation between plasma Lp(a) concentration and t-PA antigen was tested (Fig. 4). As shown in the Fig., a negative correlation between these two parameters was noted.

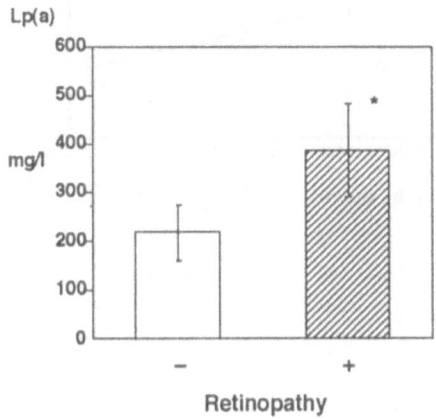

Fig.2 Plasma Lp(a) levels in diabetics with(+) or without(-) diabetic retinopathy.
* p< 0.05

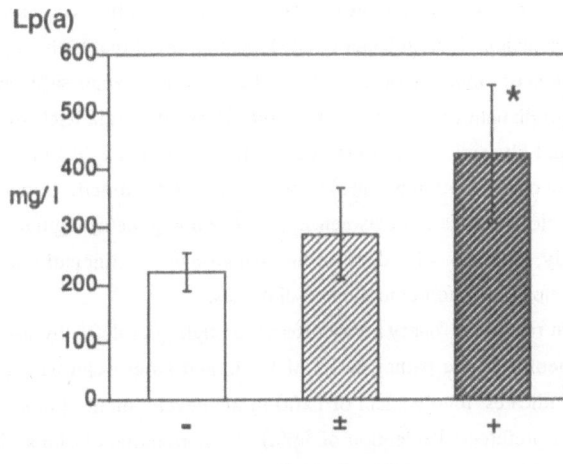

Fig.3 Plasma Lp(a) levels in diabetics with or without diabetic nephropathy. Overt nephropathy(dark column) is diagnosed by contineous albuminuria more than 300 mg/g.Cr. Albuminuria less than 300 mg/g.Cr is defined as microalbuminuria(shaded column). * p<0.05

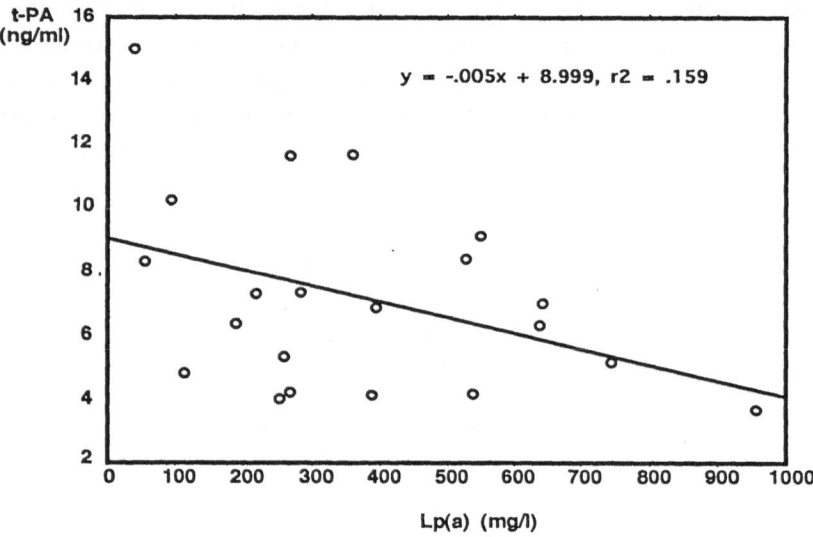

Fig.4 Correlation between plasma Lp(a) and t-PA concentration.

DISCUSSION

The present study demonstrates that plasma Lp(a) levels in diabetics is higher than in non-diabetics, and diabetics with coronary artery disease exhibit higher Lp(a) concentration compared to diabetics without coronary artery occlusion. Schernthaner et al(7) also reported that high Lp(a) concentration (above 20mg/dl) can be seen more frequently in diabetics than normal subjects. The reason for the elevation of Lp(a) in diabetics in still of postulation. Hyperglycemia itself may increase plasma Lp(a) concentration and glycemic control may reduce increased levels of Lp(a) in diabetics(8). However, because plasma concentration of Lp(a) is genetically determined, elevated Lp(a) in diabetics may be result of a close correlation between Lp(a) gene and gene(s) regulating susceptibility to diabetes. In this study, diabetics with diabetic microangiopathy, retinopathy and nephropathy, had high Lp(a) level compared to non-complicated diabetics.

Diabetes mellitus is a well-known risk for coronary artery disease, though its role(s), by itself or with concomitant hyperlipoproteinemia, in the pathogenesis of facilitated atherosclerosis remains unsolved question. Evidences indicates involvement of La(a) in the development of atherosclerosis. There is direct evidence for preferential retention of Lp(a) in atheromatous lesions via fibrin binding, in contrast to other apo-B-containing LDL(5). Macrophage uptake of Lp(a) could also be facilitated by binding of this unique lipoprotein to the endothelial cell surface via plasminogen receptor(6). The fact that Lp(a) negatively correlared with t-PA antigen level may propose an interpretation for Lp(a)'s involvement in diabetic angiopathy, thrombogenesis and atherogenesis.

REFERENCES

1. Dahlen G., Guyton JR., Atter M., Farmer JA., Kautz JA. and Gotto AM.Jr (1974) Association of levels of lipoprotein Lp(a), plasma lipids, and other lipoproteins with coronary artery disease documented by angiography. Circulation 74; 758-765.

2. McLean, JW, Tomlinson, JE, Kuang, WJ, Eaton, EY, Fless, GM and Lawn, RM.(1987) cDNA sequence of human apoprotein (a) is homologous to plasminogen. Nature 300;132-137.

3. Morrisett JD, Guyton JR, Gaubatz JW, and Gotto AM.(1987)Lipoprotein(a):structure, metabolism and epidemiology. In: Gotto AM Jr, Ed. Plasma lipoproteins, Amsterdam: Elsevier Science,pp 129-152

4. Takada, A., Shizume, K., Ozawa, T., Takahashi, S. and Takada, T. (1986) Characterization of various antibodies against tissue plasminogen activator using highly sensitive enzyme immunoassay. Thromb Res. 42; 63-72.

5. Smith EB. and Cohran S. (1990) Factors influencing the accumulation in fibrous plaques of lipid derived from low density lipoprotein II: preferential immobilization of lipoprotein(a)(Lp(a)). Atherosclerosis 84:173-181.

6. Gonzalez-Gronow M.,Edelberg JM. and Pizzo SM. (1989) Further characterization of the cellular plasminogen binding sites:evidence that plasminogen 2 and lipoprotein(a) compete for the same site. Biochemistry 28:2374-2377.

7. Schernthaner, G., Kostner, GM., Dieplinger, R. and Mülhauser, I. (1983) Apolipoproteins(A-I, A-II, Lp(a) lipoprotein and lecithin: Cholesterol acyltransferase activity in diabetes mellitus. Atherosclerosis 49:277-293.

8. Bruckert E., Davidoff P., Grimaldi A., Truffrt J., Giral P., Doumth R., Thervet F. and De Gennes J.L.(1990) Increased serum levels of lipoprotein(a) in diabetes mellitus and their reduction with glycemic control. JAMA 263(1);35-36.

EFFECT OF ETHANOL ON BLOOD SEROTONERGIC MECHANISMS

W. BUCZKO, B. MALINOWSKA, E. CHABIELSKA, D. PAWLAK

Department of Pharmacodynamics, Medical School, Mickiewicza str 2c, 15-230 Białystok, Poland.

INTRODUCTION

Ethanol may lead to the development of hypertension by central or pheripheral mechanisms.[1,3] One of the most important peripheral mechanisms responsible for the development of hypertension is the ethanol effect on blood platelet-vessel wall interaction. In clinical investigations on ethanol-dependent patients and in experiments performed on rats receiving ethanol for a long period it has been shown that the blood vessels are morc sensitive to endogenous vasoconstrictive substances.[4,5]

Serotonin (5-HT) is known to have many pharmacological effects on the peripheral cardiovascular system[6,8] and it is suggested that this amine is involved in the pathogenesis of hypertension.[9]

The aim of the present study was to determine whether the serotonergic mechanisms may play any role in the effect of ethanol on the circulatory system.

MATERIALS AND METHODS

The study was performed on male Wistar rats (200-300). Ethanol   2.0 g/kg was administrated via gastric intubation 30 min. before experiment was begun. In chronic experiments rats received ethanol 6 g/kg daily for 4 weeks. Rats were anaesthetized with phentobarbital 75.0 mg/kg intraperitoneally. Some of the animals were pithed according to the method described by Shilples and Tilden.[11] The right carotid artery was cannulated with a polyethylene tube and the blood pressure was measured via pressor transducer. To record the heart rate (HR) the subscutaneous needle electrodes were connected to the monitor. The isolated blood vessels preparation were performed according to Nicholas.[11]

RESULTS

In anaesthetized rats ethanol (2 g/kg) decreased the blood pressure and increased the heart rate (HR) 30 min. after its administration (Table 1). In pithed rats HR decreased both in the controls and experimental groups.

Table 1. Influence of ethanol (2 g/kg p.o.) on blood pressure and heart rate.

	Value of RR (mmHg)	RR 30 min.	Value of HR (beats/min.)	HR 30 min.
anaesthetized n = 10				
control	105.3 ± 18.0	2.6 ± 6.1	390.0 ± 38.7	5.9 ± 18.2
ethanol	101.3 ± 11.4	-15.0 ± 4.5**	387.0 ± 33.6	37.4 ± 22.2**
pithed n = 8				
control	52.5 ± 5.7	- 0.2 ± 1.9	331.7 ± 16.7	-11.7 ± 18.9
ethanol	46.6 ± 3.4	- 4.6 ± 5.1°	320.3 ± 15.5	-10.7 ± 22.8°

The significance of the difference from the controls $**p < 0.01$ and between anaesthetized and pithed groups $°p < 0.05$, $°°p < 0.01$.

As shown in Table 2 an ethanol concentration of 0.03M did not affect the smooth muscle of isolated tail artery and aorta. An alcohol concentration of 0.1 and 0.3M caused only a slight constriction. The maximal constriction, set at 100% was induced by ethanol in a very high concentration (10.0M).

In pithed rats serotonin produced a dose-dependent pressor effect (Table 3). The action of serotonin was decreased in a dose-dependent manner in rats pretreated with ethanol, but it was not changed in animals chronically treated with alcohol.

Similar results were obtained in vitro on the isolated rat tail artery (Table 4).

Table 2. The influence of ethanol on vasoconstrictor response of rat tail artery and aorta.

concentration of ethanol (M)	% of maximal response n = 10	
	tail artery	aorta
0.03	0.0 + 1.1	0.0 + 1.2
0.1	1.1 + 1.6	4.4 + 4.7
0.3	6.5 + 3.4	10.3 + 5.4
1.0	26.5 + 19.5	48.3 + 21.2
3.0	77.7 + 28.1	78.7 + 19.4
10.0	100.0 + 9.5	100.0 + 14.0

Table 3. Mean blood pressure (mmHg) elicited by intravenous administration of 5HT in pithed rats pretreated with ethanol.

dose of 5HT (µg/kg)	dose of ethanol (g/kg)			
	0 (n=9)	1 (n=6)	2 (n=9)	6 / 2 weeks (n=6)
3	7.6 + 4.3	6.2 + 2.0	5.5 + 1.4	5.5 + 2.8
10	12.1 + 9.0	11.5 + 2.7	7.0 + 2.3	9.7 + 5.4
30	22.7 + 9.8	26.8 + 7.6	15.7 + 6.2	33.3 + 13.9
100	66.7 +24.7	70.7 + 7.3	37.0 +10.6**	86.7 + 14.4
300	90.2 +20.2	70.1 + 3.5*	58.0 +12.7**	101.7 + 11.7
1000	92.9 +19.5	72.0 + 5.4*	67.0 + 5.7**	90.0 + 12.4

Significance of the difference from the control *$p < 0.05$, **$p < 0.01$.

Table 4. The vasoconstrictor response of rat tail artery to 5HT in the presence of ethanol.

ethanol (M)	% maximal response of controls			
	5HT 10^{-7}M	5HT 10^{-6}M	5HT 10^{-5}M	5HT 10^{-4}M
0	6.3 + 4.0	30.2 + 10.1	79.6 + 8.2	100.0 + 6.6
0.03	13.4 + 7.3*	42.3 + 16.3	90.4 + 14.3	104.3 + 14.8
0.1	6.8 + 5.9	29.9 + 11.7	66.4 + 3.1*	88.2 +11.6*
0.3	10.4 + 6.1	29.4 + 7.6	64.1 + 12.4**	76.1 + 11.1***

Significance of the difference from the control *$p < 0.05$, **$p < 0.01$, ***$p < 0.001$.

DISCUSSION

In the present study ethanol (2g/kg) caused a decrease in the mean blood pressure
and increased the HR in anaesthetized rats. The increase in HR in these animals is
in accordance with results obtained by other investigators.[1,12] The absence of an
ethanol effect on HR and blood pressure in pithed animals suggest the involvement
of the nervous system in the action of ethanol. So, the central nervous system may
be responsible for the increase in heart rate and the early phase of hypotension
caused by acute administration of alcohol.

The present results demonstrate that ethanol in a dose-dependent manner contracts
isolated rat tail artery and aorta. Our results confirm the data of others.[13-15]
It should be mentioned that only a slight contraction of isolated blood vessels
takes place only at high ethanol concentrations and marked response is produced by
concentrations that would be lethal.

One of the endogenous substances involved in the regulation of the cardiovascular
system is serotonin.[6] This amine may produce a strong contraction of smooth muscles
of blood vessels [8] and influences on the action of other agents causing vasoconstriction.[16] In our experiments we used the laboratory model of pithed rats. This
allowed us to study only the peripheral influence of ethanol on the vasopressor
action of 5HT. In pithed rats serotonin produced a pure pressor response. The
results of the experiments described in the present paper show that the pressor
response to serotonin was reduced in a dose-dependent manner by acute ethanol
administration. Chronic ethanol administration did not change the vasopressor effect
of serotonin.

We also showed that lower concentrations of ethanol augmented and higher depressed the contractive action of serotonin. Only few data concerning the effect of
ethanol on vascular effects of 5HT have been published. Toda et al.[4] have reported

that 0.1-0.3 M ethanol increased the serotonin-induced contractive response of isolated mesenteric artery of dog and rat. On the other hand Altura and Altura[4] demonstrated that high ethanol concentration inhibit contractive action of 5HT on human umbilical artery, and Dalske[17] found no effect of 0.05 M ethanol on serotonin- -induced contraction of bovine cerebral vessels. Our present results concerning isolated rat tail artery are in line with those findings. Our experiments show that ethanol significantly reduced the vasoconstrictor effect of serotonin. This action of ethanol was observed only in very high ethanol concentrations. On the other hand ethanol increased the potentiating action of serotonin on blood platelets.[18]

So, we can conclude that, among other mechanisms, the peripheral serotonergic system plays an important role in the effects of ethanol on circulation.

REFERENCES

1. Abdel-Rahman A.R.A. (1987) Acute effects of ethanol on baroreceptor reflex control of heart rate and on pressor and on depressor responsiveness in rats. Can. J. Physiol. Pharmacol. 65: 34-49.

2. Arkwirght P.D., Beilin L.J., Rouse J., Armstrong B., Vandongen R. (1981) Alcohol and blood pressure in a working population. Clin. Exp. Pharmacol. Physiol. 8: 451-454.

3. Stokes G.S. (1982) Hypertension and alcohol: Is there a link? J. Chrom. Dis. 35: 759-763.

4. Altura B.M., Altura B.T. (1983) Peripheral vascular actions of ethanol and its interaction with neurohumoral substances. Neurobehav. Toxicol. Teratol. 5: 211-220.

5. Karanian J.W., Salem N. (1986) Effect of acute and chronic ethanol exposure on the response of rat aorta to a thromboxane mimic, V 46619. Alcohol Clin. Exp. Res. 10: 171-176.

6. Göthert M. (1986) Serotonin receptors in the circulatory system. Prog. Pharmacol. 6: 155-172.

7. Marwood J.H., Stokes G.S. (1984) Review article. Serotonin (5-HT) and its antagonists: Involvement in its cardiovascular system. Clin. Exp. Pharmacol. Physiol. 11: 439-451.

8. Van Nueten J.M., Janssens W.J., Vanhoutte P.M. (1985) Serotonin and vascular reactivity. Pharmacol. res. Commun. 17: 585-608.

9. Houston D.S., Vanhoutte P.M. (1986) Serotonin and the vascular system. Role in health and disease, and implications for therapy. Drugs. 31: 149-163.

10. Shilpley R.E., Tilden J.H. (1947) A pithed rat preparation suitable for assaying pressor substances. Proc. Soc. Exp. med. 64: 453-458.

11. Nicholas T.E. (1969) A perfused tail artery preparation from the rat. J. Pharm. Pharmacol. 21: 826-832.

12. Sparrow M.G., Roggendorf H., Vogel W.H. (1987) Effect of ethanol on heart rate and blood pressure in nonstressed and stressed rats. Life Sci. 40: 2551-2559.

13. Altura B.M., Edgarian H. (1976) Ethanol-prostaglandin interactions in contraction of vascular smooth muscle. proc. Soc. Exp. Biol. Med. 152: 334-336.

14. Altura B.M., Edgarian H., Altura B.T. (1975) Differental effects of ethanol and mannitol on contraction of arterial smooth muscle. J. Pharmacol. Exp. Ther. 197: 352-361.

15. Toda N., Konishi M., Miyazaki M., Komura S. (1983) The effects of ethanol and acetylaldehyde on dog arterial smooth muscle. J. Stud. Alcohol. 44: 1-16.

16. Van Nueten J.M. (1985) Serotonin and the blood vessel wall. J. Cardiovasc. Pharmacol. 7: 549-551.

17. Dalske H.F. (1975) Effects of ethanol and its interaction with vasoactive agents in bovine vascular smooth muscle. Arch. Int. Pharmacodyn. 218: 54-65.

18. Chabielska E., Malinowska B., Buczko W. (1990) Influence of ethanol and serotonin on rat platelet aggregation. Pharmacology 40: 288-292.

SNAKE VENOM PROTEINS AFFECTING PLATELET FUNCTIONS

C.M. TENG AND T.F. HUANG
Pharmacological Institute, College of Medicine, National Taiwan University, Taipei, Taiwan.

Snake venoms are complex mixture of proteins with various biological activities, such as neurotoxins, hemorrhagins, coagulants, anticoagulants and cardiotoxins[1]. The study of these venom proteins leads to our understanding of snakebite symptoms and also helps their rational treatments. In the last decade, many stimulants and inhibitors of platelet activation have been isolated from the snake venoms in our laboratory. In this paper, the effects of these venom proteins on platelet functions are reviewed. According to their biochemical properties and actions on platelets, they are classified into seven groups. The venom proteins of the first three groups accelerate or activate platelet aggregation, while those of the other four groups inhibit this process.

1. Thrombin-like Enzymes

Thrombin-like enzymes are found in most venoms of snakes from Crotalidae family[2,3]. Although they coagulate fibrinogen, their properties are found to be different from those of thrombin. Most thrombin-like enzymes release preferentially fibrinopeptide A, fail to activate Factor XIII and their coagulant activities are not inhibited by heparin and antithrombin III[4].

Three thrombin-like enzymes have been purified and their effect on fibrinogen and platelets were compared e.g. thrombocytin and batroxobin from *B. atrox* venom[5] and acutin from *A. acutus* venom[6]. The clotting activity of these venom proteins decreases in the following order: batroxobin > acutin > thrombocytin. However, their stimulatory effects on platelets are in the opposite order. A detailed comparison of the dose-response curve reveals that the aggregating activity of thrombin is 10^2, 10^4 and 10^5 times more potent than those of thrombocytin, acutin and batroxobin, respectively. Platelet-activating potency of the thrombin-like enzymes is correlated with their effectiveness on the retractility and elasticity of the clots[7].

The aggregating activity of thrombin-like enzymes could not be inhibited by indomethacin or platelet-activating factor (PAF) antagonists. Only small amount of thromboxane B_2 formed during platelet activation. However, ADP-scavenging system, creatine phosphate/creatine phosphokinase inhibited completely the aggregation caused by thrombin-like enzymes but not that by thrombin. Unlike thrombin, ADP-release mechanism induced by thrombin-like enzymes is dependent on the extracellular calcium[7]. Like thrombin, thrombin-like enzymes are serine

proteases, and their enzymatic and clotting activities can be inhibited by diisopropyl fluorophosphate, phenylmethanesulfonyl fluoride or tosyl-lysine chloromethylketone[2].

2. Noncoagulant, Nonenzymatic Inducers

Many noncoagulant platelet-aggregating proteins have been purified, and shown to be devoid of any recognized enzymatic activity found in the crude venoms. These include trimucytin isolated from *T. mucrosquamatus* venom[8] and triwaglerin isolated from *T. wagleri* venom[9]. These venom inducers cause aggregation, not agglutination, of platelets, because they do not clump the formaldehyde-fixed platelets. Their aggregating actions are calcium-dependent and accompanied with release reaction and thromboxane formation. Although indomethacin inhibits the release reaction and thromboxane formation, however, the aggregation is apparently not affected by indomethacin. This class of venom inducers cause "novel" aggregation which is thromboxane A_2-, ADP- and PAF-independent. The aggregation induced by triwaglerin is inhibited completely by mepacrine, imipramine and forskolin, and markedly by tetracaine and sodium nitroprusside. It was suggested that triwaglerin induced platelet aggregation possibly through phospholipase C-phosphoinositide mechanism[9].

Trimucytin caused morphological changes similar to thrombin or collagen even its action was thromboxane- and release- independent. Platelets exposed to trimucytin lost their discoid shape and develope irregular projections, which are devoid of organelles. Microtubules form a collar around the organelles which tend to shift toward the platelet center. Most α-granules and dense-bodies finally disappear and elements of tubular systems remain in the ballooned cytoplasm in the periphery of the aggregate[10].

3. Membrane-active Polypeptides

Membrane-active polypeptides are basic and devoid of enzymatic activity, existing in the snake venoms of Elapidae family. They induce a variety of pharmacological effects, such as muscle contracture, direct hemolysis and cytotoxic action. Cardiotoxin, a membrane-active polypeptide isolated from *N. n. atra* venom, potentiates platelet aggregation induced by ADP, thrombin, collagen and venom phospholipases A_2. However, cardiotoxin causes cell lysis at high concentrations. The potentiation of aggregation and increase of thromboxane B_2 formation are blocked by indomethacin and calcium of high (5 mM) or low (0.05 mM) concentration. Cardiotoxin does not potentiate thrombin-induced aggregation of p-bromophenacyl bromide-modified platelets, indicating activation of endogenous phospholipase A_2 is involved with its action. Arachidonic acid-induced platelet aggregation is not affected. So, cardiotoxin may augment the calcium-flux during the activation of platelets by the aggregation inducers and subsequently increase the activation of endogenous phospholipase A_2[11].

4. Phospholipase A_2 Enzymes

Platelet membrane phospholipids play important roles in blood coagulation and platelet aggregation. Phospholipases A_2 exist in almost every kind of snake venoms. Platelet

membrane phospholipids are easily assessible to venom phospholipases A_2. These enzymes also attack tissue thromboplastin and possess anticoagulant action[12-14].

Most venom phospholipases A_2, e.g. from the venoms of *N. n. atra, T. mucrosquamatus* and *V. russelli formosensis*[15] exhibit biphasic effects on platelet aggregation - an immediate aggregating effect and a delayed inhibitory effect. The aggregation and ATP release can be completely inhibited by EDTA, indomethacin, mepacrine and prostaglandin E_1. Venom phospholipases A_2 cause thromboxane formation which is inhibited by EDTA or indomethacin. They also release free fatty acids from synthetic phosphatidylcholine and intact platelets. p-Bromophenacyl bromide-modified phospholipases A_2 lose their phospholipid-hydrolyzing and platelet-activating activities. All the above aggregation characteristics are similar to those produced by exogenous addition of arachidonic acid, including a bell-shape dose-response relationship[14,16,17].

In the presence of bovine serum albumin, venom phospholipases A_2 can not induce aggregation, and their pretreatment with platelets will inhibit the aggregation induced by arachidonic acid or collagen. The antiplatelet effect is dependent on the incubation time of phospholipases A_2 with platelets. Lysophosphatidylcholine and arachidonic acid can mimic this inhibitory effect. Thus, venom phospholipases A_2 show stimulating effect on platelet aggregation by virtue of their enzymatic hydrolysis of platelet phospholipids resulting in arachidonic acid release and formation of thromboxane A_2. On the other hand, the cleaved products, lysophospholipid, arachidonic acid or its metabolite(s) (probably via lipoxygenase pathway) may be responsible for the antiplatelet action[16,17].

5. Fibrinogenolytic Enzymes

Fibrinogenolytic enzymes are first purified from *T. mucrosquamatus* venom and are classified into α- and β-fibrinogenases (EC 3,4,21,5) according to their specificity of the digestion on the polypeptide chains of fibrinogen molecule[18]. Thus the actions of both venom fibrinogenases are more specific than those of trypsin and plasmin. Since then, many fibrinogenases were isolated and classified according to the above criteria. α-Fibrinogenases, but not β-fibrinogenases, inhibit the aggregation of washed rabbit platelets induced by ADP. The inhibition of platelet aggregation by α-fibrinogenases is dependent on the concentration of the enzymes and incubation time with platelets, indicating that an enzymatic aciton is involved.

The importance of α(A) and γ, but not β(B) chains of the fibrinogen molecule in the interaction of fibrinogen-platelet has been reported by Niewiarowski *et al.*[19]. Lack of the antiplatelet effect of β- fibrinogenases of snake venoms is then easily explained. The inhibitory effect of α-fibrinogenases on platelet aggregation is much less evident in the preparation of platelet-rich plasma. This may be due to the presence of the protease inhibitors and excess of fibrinogen in the plasma.

Recently, a novel α-fibrinogenase (kistomin), isolated from *C. rhodostoma* venom has been shown to inhibit platelet aggregation induced by ristocetin or low concentration of thrombin. This antiplatelet activity results from the selective cleaving activity on platelet membrane

glycoprotein Ib[20]. Another α-fibrinogenase (caprotease), isolated from *C. atrox* venom, can inhibit collagen-induced platelet aggregation probably by cleavage of platelet membrane glycoprotein Ia/IIa complex (unpublished data).

6. ADP-splitting Enzymes

Although biochemical characterization of ADPase or 5'-nucleotidases of snake venoms have been reported by many investigators, only those from *T. gramineus*[21] and *A. acutus*[22] venoms have been shown to inhibit platelet aggregation. The 5'-nucleotidase of *T. gramineus* is a metalloprotein and its enzyme activity is inhibited by EDTA. Both Zn^{++} and Co^{++} reverse the inhibitory effect of EDTA. The ADPase of *A. acutus* venom is a single peptide chain with ADP-hydrolyzing activity of 4.3 μmol/min/mg[22].

5'-Nucleotidase or ADPase inhibits the aggregation of platelet-rich plasma induced by ADP, arachidonic acid, collagen and ionophore A23187. Less inhibition is observed in platelet suspension. The removal of ADP, which is released by the aggregation inducers, and subsequent accumulation of adenosine are responsible for the inhibitory effect of the venom 5'-nucleotidase on platelet aggregation[21,22].

7. Fibrinogen-receptor Antagonists

Several potent platelet aggregation inhibitors have been purified from snake venoms e.g. trigramin from *T. gramineus*[23-25], halysin from *A. halys*[26,27], and rhodostomin from *C. rhodostoma*[28]. These trigramin-like peptides (TLPs) inhibited aggregation nonenzymatically induced by a variety of aggregation agonists including ADP, epinephrine, sodium arachidonate, collagen, thrombin and calcium ionophore A23187. They exert little effect on the initial shape change and release reaction of the activated platelets. Sequence analysis showed that they are cysteine-rich, single chain polypeptides with Arg-Gly-Asp (RGD) sequence near their carboxyl terminus[29]. Tripeptide RGD is thought to be the recognition site of many adhesive proteins, such as fibrinogen, vitronectin, fibronectin and von Willebrand factor, for their binding activity toward their respective receptors[30]. Trigramin blocks fibrinogen-induced aggregation and ^{125}I-fibrinogen binding to ADP-stimulated platelets, and inhibits fibrinogen-induced aggregation of α-chymotrypsin-treated platelets. ^{125}I-trigramin binds to unstimulated and ADP-stimulated platelets with a Kd value of $2x10^{-8}$ M (in the ADP-stimulated platelets). RGDS and $7E_3$, monoclonal antibody against glycoprotein IIb/IIIa complex, block ^{125}I-fibrinogen binding as well as ^{125}I-trigramin binding[29]. All these evidences suggest that trigramin preferentially binds to fibrinogen receptor associated with glycoprotein IIb/IIIa complex on platelet membrane, leading to blockade of fibrinogen binding to ADP-stimulated platelets and subsequently blockade of platelet aggregation. Direct evidence came from the result that trigramin would not bind to the thrombasthenic platelets in which the content of glycoprotein IIb/IIIa complex was less than 5% of normal platelets[29].

In addition to the antiplatelet activity, TLPs inhibit the binding of von Willebrand factor to thrombin-stimulated platelets[29], inhibit the adhesion of melanoma cells to fibrinogen- and fibronectin-coated plates[31,32] and prevent the platelet plug formation of the severed hamster mesentery arteries[33].   They are useful tools for the study of the interaction of fibrinogen and fibrinogen receptor associated with glycoprotein IIb/IIIa complex of the platelets. The elucidation of the structure-activity relationship would lead to the development of the potential antithrombotic agents in the near future.

Acknowledgement:

This work was supported by a research grants of the National Science Council of the Republic of China (NSC78-0412-B002-60).

REFERENCES

1. Ouyang C, Teng CM, Huang TF (1987) Asia Pacific J Pharmacol 2: 169-176
2. Seegers WH, Ouyang C (1979) Lee CY (ed) Snake Venoms, Handbook of Experimental Pharmacology, Springer-Verlag, Berlin-Heidelberg-New York, 52: 684-750
3. Stocker KF, Meier J (1988) In: Pirkle H, Markland FS Jr (ed) Hemostasis and Animal Venoms, Marcel Dekker, New Yrok, pp 67-84
4. Stocker K, Fischer H, Meier J (1982) Toxicon 20: 265-273
5. Teng CM, Ko FN (1988) J Biomed Lab Sci 1: 98-107
6. Ouyang C, Teng CM (1978) Toxicon 16: 583-593
7. Teng CM, Ko FN (1988) Thrombos Hemostas 59: 304-309
8. Ouyang C, Wang JP, Teng CM (1980) Biochim Biophys Acta 630: 246-253
9. Teng CM, Huang ML, Ouyang C, Huang TF (1989) Biochem Biophys Acta 992: 258-264
10. Teng CM, Liao KK, Wang JP, Lin HS, Ouyang C (1981) Toxicon 19: 121-130
11. Teng CM, Jy W, Ouyang C (1984) Toxicon 22: 463-470
12. Ouyang C, Teng CM, Chen YC, Lin SC (1978) Biochim Biophy Acta 541: 394-407
13. Ouyang C, Jy W, Zan YP, Teng CM (1981) Toxicon 19: 113-120
14. Teng CM, Chen YH, Ouyang C (1985) Sem Thromb Hemosta 11: 369-374
15. Teng CM, Chen YH, Ouyang C (1984) Biochim Biophys Acta 786: 204-212
16. Teng CM, Kuo YP, Lee LG, Ouyang C (1986) Thromb Res 44: 875-886
17. Teng CM, Chen YH, Ouyang C (1984) Biochim Biophys Acta 772: 393-402
18. Ouyang C, Teng CM (1976) Biochim Biophys Acta 420: 298-308
19. Niewiarowski S, Budzynski AZ, Morinelli TA, Budzynski TM, Stewart GJ (1981) J Biol Chem 256: 917-925
20. Huang TF, Chang MC (1991) In: Sixth Southeast Asian/Western Pacific Regional Meeting of Pharmacologists, Hong Kong, Abstract 3154
21. Ouyang C, Huang TF (1983) Toxicon 21: 491-501
22. Ouyang C, Huang TF (1986) Toxicon 24: 1099-1106
23. Huang TF, Holt JC, Lukasiewcz H, Niewiarowski S (1987) J Biol Chem 1987 262: 16157-16163
24. Ouyang C, Huang TF (1983) Biochim Biophys Acta 757: 332-341
25. Huang TF, Ouyang C (1984) Thromb Res 33: 124-138

26. Ouyang C, Yeh HI, Huang TF (1983) Toxicon 21: 797-804

27. Huang TF, Yeh HI, Ouyang C (1984) Toxicon 22: 243-251

28. Huang TF, Wu YJ, Ouyang C (1987) Biochim Biophys Acta 925: 248-257

29. Huang TF, Holt JC, Kirby EP, Niewiarowski S (1989) Biochem 28: 661-666

30. Hynes RO (1987) Cell 48: 549-554

31. Knudsen KA, Tuszynski GP, Huang TF, Niewiarowski S (1989) Exp Cell Res 179: 42-49

32. Niewiarowski S, Rucinski B, Huang TF, Williams JA, Holt JC, Tuszynski GP, Knudsen KA (1989) Gordon Conference on Fibronectin, Oxnard, CA

33. Cook JC, Huang TF, Rucinski B, Stryzewski M, Tuma RF, Williams JA, Niewiarowski S (1989) Am J Physiol 256: H1038-1043

ALTERED PLATELET RESPONSES TO THROMBOXANE A_2 IN PATIENTS WITH MYELO-PROLIFERATIVE DISORDERS

MINORU OKUMA[1], FUMITAKA USHIKUBI[2], TAKAFUMI ISHIBASHI[1], SHU NARUMIYA[2], KENJIRO TOMO[1] AND HIROSHI TAKAYAMA[1]
[1]The First Division, Department of Internal Medicine, and [2]Department of Pharmacology, Faculty of Medicine, Kyoto University, Sakyo-ku, Kyoto, Japan

INTRODUCTION

Thromboxane A_2 (TXA$_2$), a major arachidonic acid (AA) metabolite in platelets, is a potent stimulator of platelet functions and constrictor of vascular and respiratory smooth muscles [1,2]. Because of such potent biological activities and its formation in response to various stimuli, TXA$_2$ has been implicated as a mediator in many pathophysiological conditions including myocardial infarction, stroke and anaphylaxis [2,3]. Furthermore, platelet dysfunctions due to defective TXA$_2$ synthesis or subnormal responses to TXA$_2$ are associated with bleeding tendency [4].

Myeloproliferative disorders (MPD) including polycythemia vera (PV), essential thrombocythemia (ET), chronic myeloid leukemia (CML) and myelofibrosis (MF) are frequently complicated with bleeding and/or thrombotic tendencies [5]. Although various kinds of platelet abnormalities have been reported, mechanisms of such tendencies remain to be elucidated [5,6]. We found not only increased platelet response to TXA$_2$ in most of the MPD patients [6], but also defective one in very rare cases of the patient with this disorder [7,8]. Thus, mechanisms of these altered platelet responses to TXA$_2$ were studied.

MATERIALS AND METHODS

Patients Seventy-six patients with MPD including 12 with ET, 25 with PV, 33 with CML and 6 with MF were routinely studied for platelet aggregation. One [7] of the two patients with subnormal platelet responses to TXA$_2$ [7,8] was employed for this study to analyze defective signal transduction mechanism through the platelet TXA$_2$ receptor. Consecutive 11 patients exclusive of these 2 patients were employed for another part of this study: 6 with PV, 4 with ET, 1 with CML; age between 39 and 76 years old; 4 males, 7 females.
Controls Healthy adult donors who had taken no drugs for at least two weeks were chosen.

All studies were done after informed consent was obtained.
Materials ONO3708 (9,11-dimethylmethano-11,12-methano-16-phenyl-13,14-dihydro-13-aza-15-β-ω-tetranor-TXA$_2$) was a generous gift from Ono Pharmaceutical Co., Osaka, Japan. [γ-^{32}P] guanosine trisphosphate (3000 Ci/mmol) was obtained from New England Nuclear, Boston, MA. AMP-PNP (β-γ-imidoadenine-5'-triphosphate) was obtained from Sigma Chemical Co., St. Louis, MO. All other materials were obtained from the same sources as described previously [7,8,9].
Platelet function studies Unless otherwise specified, the collection of blood, isolation of platelet-rich plasma (PRP) and platelet-poor plasma (PPP) were done as previously reported [8]. Platelet aggrega-

Key words: Platelet, thromboxane A_2, myeloproliferative disorder, receptor

tion was also studied as previously reported [8], and the concentration of STA$_2$ (a stable TXA$_2$ mimetic [8]) to induce half maximal increase of light transmittance in STA$_2$-induced platelet aggregation (E C$_{50}$) was determined by PRP's whose platelet counts had been adjusted to 300 x 10^3/μl by autologous PPP.

Down regulation of the TXA$_2$ receptor   Blood was obtained in one tenth volume of 3.8 % trisodium citrate and 10 μM indomethacin (final concentration). The blood was centrifuged for 10 min at 160 x g and PRP was collected.   PRP was routinely incubated with 2 μM STA$_2$, 2 μM ONO3708 (a TXA$_2$ receptor antagonist), 0.5 μM TPA (12-O-tetradecanoyl-phorbol-13-acetate; a protein kinase C activator) or vehicle (control) in the presence of 7.7 mM EDTA (final concentration) at room temperature for 20 hr.   Washed platelets were prepared as described previously [9].   TXA$_2$ receptors on washed platelets were analyzed by binding studies using a TXA$_2$ antagonist [^{3}H]S-145 as a radioligand [8,9].

Measurement of GTPase activity   Platelet high-affinity GTPase activity was assessed for platelet membranes essentially by the method of Cassel et al [10] with minor modifications. Platelet membranes were prepared as reported previously [9] from washed platelets which had been incubated with or without 1 μM STA$_2$ for 30 min at room temperature in the presence of 7.7 mM EDTA.   High-affinity GTPase activity was defined as that inhibited by the addition of 1 mM GTP in the assay. The reaction mixture (50 μl) contained 50 μg membrane protein, 3 mM MgCl$_2$, 1 mM EGTA, 0.2 mM ATP, 2 mM AMP-PNP, 1 mM phosphocreatine, 50 units creatine phosphokinase, 0.2 mM dithiothreitol, 100 mM NaCl, 0.1 μM [γ-^{32}P]GTP (20 Ci/mmol), 20 mM Tris-HCl (pH7.4) and several concentrations of STA$_2$.   After the assay mixture was incubated for 10 min at 37°C, 1 ml of ice-cold 5% Norit A in 20 mM phosphate buffer (pH7.4) was added to it.   The mixture was then centrifuged at 15,000 rpm for 5 min in Tomy Microfuge (Tomy Seiko Co., Tokyo).   Aliquots (500 μl) of the suprnatant were counted for radioactivity on a Hitachi Scintillation system.   A blank value was less than 5% of total [γ-^{32}P]GTP added.

Binding assays   Binding assays to washed platelets and platelet membranes were carried out as described previously [9].

Protein determination   Protein was determined as described previously [9].

Statistical analysis   If not otherwise indicated, all data given in the text are means (SD).   The results were analyzed by using the Student's t-test for unpaired data and by linear regression analysis where appropriate. In some studies, the difference was considered significant when the value for the patient exceeded beyond means ± 2SD of the control values.

RESULTS

    Out of the 76 MPD patients studied, only one each with CML [7] and PV [8] showed subnormal or defective platelet responses to AA and enzymatically generated TXA$_2$/prostaglandin H$_2$ (PGH$_2$).

Defective platelet response to TXA$_2$ in a PV patient with a mild bleeding tendency   We reported previously a PV patient who had hemorrhagic thrombocytopathy with platelet TXA$_2$ receptor abnormality, and defective signal transduction with normal binding activity was demonstrated [8].   In short, a 56-year old man with PV had recurrent cutaneous ecchymoses, occasional gum bleeding since 3 years and episodes of severe bleeding from hemorrhoids at 55 years old.   Bleeding time was prolonged, but coagulation tests were all within normal ranges.   All agonists except for thrombin induced defective or subnormal aggregation responses of the patient's platelets, and it was to be noted that STA$_2$ as well as enzymatically generated TXA$_2$/PGH$_2$ could not induce his platelet aggregation.   These results suggested that abnormalities of the patient's platelet included a defective response to TXA$_2$. Arachidonate metabolism, adenine nucleotide contents and TXA$_2$ binding activities of the patient's platelet were all normal. The rise in the cytoplasmic Ca ion concentration, phosphatidylinositol breakdown and 40 kD protein phosphorylation were significantly reduced in response to STA$_2$, but thrombin induced their normal responses in the patient's platelets. These results suggested that an abnormality of the patient's platelets was not in the ligand binding but in the coupling mechanism between TXA$_2$ binding and postreceptor events. Therefore, we further analyzed STA$_2$/GTP binding protein coupling by STA$_2$-stimulated GTPase assay and STA$_2$-induced down regulation of the TXA$_2$ receptor in this patient.

STA$_2$ stimulated high affinity GTPase activities in a dose-dependent manner, but the increase in the patient was significantly lower than in normal subjects and in other MPD patients, while the increases in normal subjects and other MPD patients were similar (Fig. 1). In experiments of STA$_2$-induced down regulation of the TXA$_2$ receptor, the receptor density of normal platelets decreased to about a half of controls, while that of the patient's platelets decreased to only 91% of controls. On the other hand, when normal platelets were incubated with ONO3708 (a TXA$_2$ receptor antagonist) or with TPA (a protein kinase C activator), the receptor density did not change (Fig. 2). Thus, only STA$_2$ induced down regulation of the receptor in normal platelets, but this effect of STA$_2$ was significantly decreased in the patient's platelets.

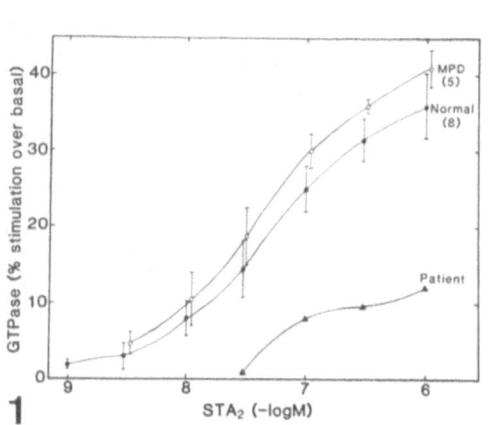

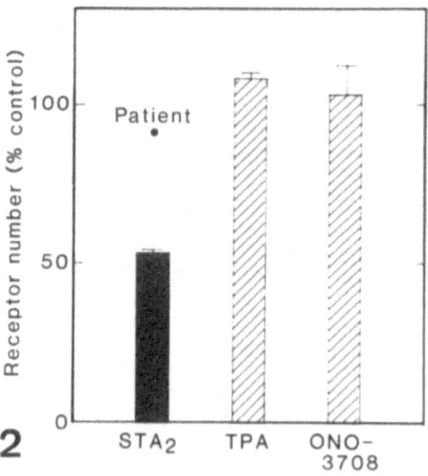

Fig. 1. GTPase activities of STA$_2$-stimulated platelet membranes. The values for normal subjects (●) and MPD patients (O) represent the means ± SD. ( ), N; ▲, the patient

Fig. 2. Down regulation of the TXA$_2$ receptor. Numbers of the TXA$_2$ receptor of normal platelets after the incubation with STA$_2$, TPA or ONO3708 (columns; means ± SD, n = 4) and of the patient's platelets after the incubation with STA$_2$ (●) are shown.

Increased platelet sensitivity to TXA$_2$ and enhanced expression of the TXA$_2$ receptor When the EC$_{50}$ values for STA$_2$-induced platelet aggregation were compared between the MPD patients exclusive of those with subnormal platelet responses to TXA$_2$ [0.32 ± 0.14 μM (M ± SD, n =

11)] and normal subjects (0.81 ± 0.21 μM, n = 11), they were significantly smaller in the former than in the latter (p<0.001). This result suggests that the patient's platelets are more sensitive to TXA$_2$ than normal ones.

When we compared the expression of TXA$_2$ receptors on washed platelets and their membranes between the MPD patients and normal subjects by the binding assay of [^{3}H]S-145, Scatchard analysis of the radioligand binding to the platelet membrane revealed that Bmax values for the patients were significantly higher than those for normal subjects (p<0.001), although those values of washed platelets were similar in both groups (Fig. 3). Kd values of washed platelets and platelet membranes were comparable between the patients and normal subjects (data not shown). Fig. 4 shows correlation between the EC$_{50}$ values for STA$_2$-induced platelet aggregation and the Bmax values for [^{3}H]S-145 binding to the platelet membranes in the MPD patients and normal subjects. Significantly negative correlation was found between them with a correlation coefficient of -0.8 and the p value of 0.005. Out of the 11 MPD patients investigated, 3 had histories of thrombotic tendencies and 2 had those of bleeding tendency, while the rest of the patients had neither of them.

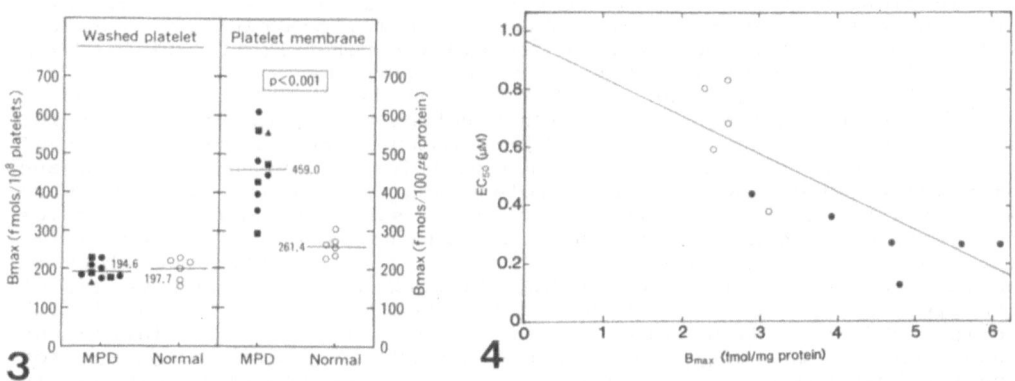

Fig. 3. Bmax values by Scatchard analysis of [^{3}H]S-145 binding to washed platelets and platelet membranes. ●,PV; ■,ET; ▲,CML; ○, normal; ——,mean

Fig. 4. Correlation between the EC$_{50}$ values for STA$_2$-induced aggregation and the Bmax values for [^{3}H]S-145 binding to the platelet membrane. ●, MPD patient; ○, normal subject

DISCUSSION

We found two kinds of altered platelet responses to TXA$_2$, i.e., subnormal and increased responses in MPD patients. Only 2 of 76 patients showed the subnormal response and both of them had a mild bleeding tendency [7,8]. In the present study, we further analyzed platelets from one of these two patients [8]. The patient's platelets showed defective aggregation and release reaction when stimulated by AA, TXA$_2$ and synthetic TXA$_2$ receptor agonists such as STA$_2$ and U-46619 [8]. We also reported previously that the patient's platelets had normal ligand binding activity but subnormal generation of second messengers [8].

We studied agonist-stimulated GTPase activities as a measure of the coupling of the TXA_2 receptor with its relevant GTP binding protein. Increase of STA_2-stimulated GTPase activity in the patient's platelets was significantly lower than that of normal platelets. This means that the coupling of the receptor with the GTP binding protein was defective. Although STA_2 induced down regulation of the receptor in normal platelets, this effect of the TXA_2 receptor agonist was significantly decreased in the patient's platelets. The mechanism of down regulation of the TXA_2 receptor is unknown, but phosphorylation of the receptor may be involved. TPA did not induce down regulation of the TXA_2 receptor in normal platelets, and this means that the feedback mechanism through protein kinase C does not contribute to the receptor down regulation. ONO3708 did not induce receptor down regulation. These results imply that agonist-induced conformational change of the receptor is necessary for the receptor down regulation, and this might be due to the receptor phosphorylation by a receptor kinase as in the case of β_2-adrenergic receptor [11]. It could be speculated that, in the patient's platelets, TXA_2-induced conformational change of the receptor was defective and that this led to defective coupling with GTP binding protein and/or abnormal down regulation of the receptor. Thus, it was again suggested that an abnormality of the patient's platelet was in the TXA_2 receptor itself despite its normal ligand binding activity.

Platelets from MPD patients except for very rare cases as described above were more sensitive to STA_2 than normal platelets, since EC_{50} values for STA_2-induced platelet aggregation were significantly smaller in those MPD patients than in normal subjects. To elucidate the alteration in the response to TXA_2, we analyzed TXA_2-ligand binding activity by using [^{3}H]S-145, a highly potent and specific radioligand, to the platelet TXA_2 receptor [9]. Although Scatchard analysis of this binding to the patients' washed platelets revealed normal Kd and Bmax values, Bmax values for platelet membranes of the patients were significantly higher than those for normal subjects, and Kd values for the platelet membranes of the patients and normal subjects were similar. Thus, the affinity of the patient's TXA_2 receptor seemed to be normal, but the total number of TXA_2 binding sites per unit weight of the platelet membrane was increased in the MPD patients compared with normal subjects. Furthermore, significantly negative correlation was found between the EC_{50} values for STA_2-induced platelet aggregation and the Bmax values for [^{3}H]S-145 binding to the platelet membrane in the patients and normal subjects. These results suggest that the increased maximal number of TXA_2 binding sites rather than its affinity on the platelet membrane could contribute to the enhanced platelet response to TXA_2. Cortelazzo et al reported an increased response to AA and U-46619, a PGH_2 mimetic, and resistance to inhibitory PG's in MPD patients [12], although they did not mention about defective platelet response to TXA_2 in MPD patients we found previously [7,8]. In our studies, defective platelet response to TXA_2 was associated with bleeding tendencies, while enhanced platelet sensitivity to TXA_2 was not always associated with thrombotic tendency, although such an association could be speculated. These findings on altered platelet responses to TXA_2, however, could have a clinical relevance to the hemostatic and thrombotic complications so frequently observed in MPD patients. Pathophysiological significance of the increased platelet response to TXA_2 remains to be elucidated. Molecular characterization of the TXA_2 receptor of the patient's platelet should clarify the exact nature of the alterations and help understand receptor functions.

CONCLUSION

1. We found two kinds of altered platelet responses to TXA_2 in MPD patients: defective and increased responses.
2. The patient with a defective platelet response to TXA_2 is very rare and such patients show mild bleeding tendencies. Analysis of the signal transduction pathway in platelets of one of such patients suggested an abnorality in the TXA_2 receptor itself despite its normal binding activity with the agonist.
3. Platelets of most MPD patients showed increased sensitivity to TXA_2, and this seemed to be due to the increased maximal number of TXA_2 binding sites on the platelet membrane.

ACKNOWLEDGMENT

We thank Shionogi Research Laboratory and Ono Pharmaceuticals for their generous supply of TXA_2 agonists and antagonists. This work was supported in part by a grant-in-aid for Scientific Research from the Ministry of Education, Science and Culture of Japan.

REFERENCES

1. Samuelsson B, Goldyne M, Granström E, Hamberg M, Hammerström S, Malmsten C (1978) Ann Rev Biochem 47: 997-1029

2. Granström E, Diczfalusy U, Hamberg M (1983) Prostaglandins and related substances. Elsevier Science Publishers. Amsterdam, pp45-94

3. Ogletree ML (1987) Fed Proc 46: 133-138

4. Hardisty RM, Caen JP (1987) Haemostasis and thrombosis. Churchill Livingstone. London, pp365-392

5. Schafer Al (1984) Blood 64: 1-12

6. Okuma M (1989) Acta Haematol Jpn 52: 1265-1272

7. Okuma M, Takayama H, Uchino H (1982) Br J Haematol 51: 469-477

8. Ushikubi F, Okuma M, Kanaji K, Sugiyama T, Ogorochi T, Narumiya S, Uchino H (1987) Thromb Haemost 57: 158-164

9. Ushikubi F, Nakajima M, Yamamoto M, Ohts K, Kimura Y, Okuma M, Uchino H, Narumiya S (1989) Eicosanoid 2: 21-27

10. Cassel D, Selinger Z (1977) Biochem Biophys Res Commun 77: 868-873

11. Murray R, FitzGerald GA (1989) Proc Natl Acad Sci USA 86: 124-128

12. Cortelazzo S, Galli M, Castagna D, Viero P, de Gaetano G, Barbui T (1988) Thromb Haemost 59: 73-76

Fibrinolysis

EFFECTS OF VARIOUS N-TERMINAL PEPTIDES OF GLU-PLASMINOGEN ON KEEPING ITS TIGHT CONFORMATION OF GLU-PLASMINOGEN

TETSUMEI URANO, YUMIKO TAKADA AND AKIKAZU TAKADA

Department of Physiology, Hamamatsu University School of Medicine, Hamamatsu, Shizuoka, Japan

INTRODUCTION

Glu-plasminogen (Glu-plg) is a single chain plasma glycoprotein, which is the zymogen form of a serine protease, plasmin. The conversion from the precursor to the enzyme occurs as a result of cleavage of the Arg^{560}-Val^{561} peptide bond in plasminogen splitting to the heavy chain and the light chain, by several activators, which include tissue type plasminogen activator (tPA), urokinase type plasminogen activator (uPA) and streptokinase [1]. In heavy chain of plasmin which coincides with the sequence of Glu^1-Arg^{450} of Glu-plg, there are five conformationaly similar loop domains called kringle domain which binds to many regulatory molecules like fibrin and lysine analogues. Glu-plg possesses a tight conformation in the presence of chloride ion and absence of lysine analogues, and is hardly activated by activators, whereas Lys-plasminogen (Lys-plg), which is generated by the cleavage of Lys^{76}-Lys^{77} peptide bond of Glu-plg by plasmin, has a looser conformation and is readily activatable [2~9]. The tight conformation of Glu-plg is considered to be caused by the intramolecular binding of N-terminal portion to the kringle 1 [10, 11] or the kringle 4 domain [12] through lysine binding site (LBS) or kringle 5 domain through amino-hexyl (AH) site [13] on the same molecule. Lysine analogues, therefore, could alter the conformation of Glu-plg to the looser form [3, 6, 8, 14] by dissociating the N-terminal peptides from LBS or AH site. Such looser conformation is supposed to be similar to that in the presence of fibrin. It is still not clear, however, which is the binding site between LBS or AH site and also which lysine residue in N-terminal region of Glu-plg is responsible for this intramolecular binding. In the present study we discuss about the mechanism for the intramolecular binding in Glu-plg molecule.

EFFECTS OF LYSINE AND ITS ANALOGUES ON THE ACTIVATION OF GLU-PLG.

The effects of EACA on Glu-plg activation is summarized in Fig. 1. It is shown that Glu-plg is hardly activatable in the absence of EACA and the presence of physiological concentration of chloride ion [7, 8]. In the presence of EACA, however, the activation rate is enhanced mainly by

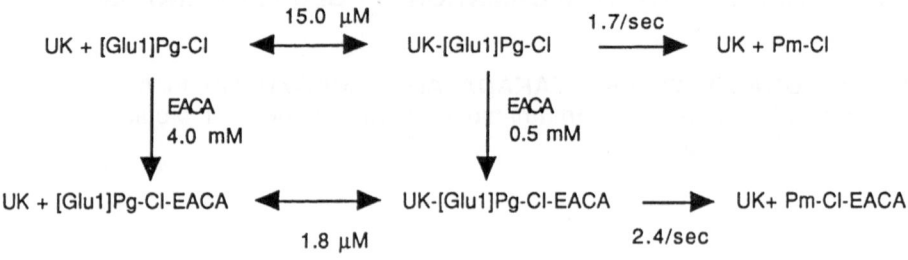

Figure 1. Effect of EACA on the activation of Glu-plg by uPA in the presence of 100 mM NaCl.

decreasing km value [8]. The mechanism of the effects of EACA are considered to change the susceptibility of Glu-plg by uPA by altering the conformation of Glu-plg molecule by dissociating the intramolecular binding of N-terminal peptide, which are shown by intrinsic fluorescent study and studies of sedimentation coefficient [6, 8].

In order to bind to LBS, lysine analogues such as EACA or tranexamic acid must have positively charged amino radical at N-terminus and negatively charged carboxyl radical at the distance close to these radicals of lysine. When lysine resides in the middle of the protein, however, negatively charged carboxyl residue is not available for the binding due to its formation of a peptide bond with next amino acid. In order for lysine residue in the middle of the protein to bind to LBS, therefore, we believe that the coexistence of amino acid with negatively charged side chain adjacent to lysine residue is prerequisite.

```
            10                      20                      30
NH2-E-P-L-D-D-Y-V-N-T-Q-G-A-S-L-F-S-V-T-K-K-Q-L-G-A-G-S-I-E-E-C-
            40                      50                      60
  A-A-K-C-E-E-D-E-E-F-T-C-R-A-F-Q-Y-H-S-K-E-Q-E-C-V-I-M-A-E-N-
            70                77
  R-K-S-S-I-I-R-M-R-D-V-V-L-F-E-K----K-V-Y-L-
```

Scheme 1. Amino acid sequence of N-terminal region of Glu-plg.

INTRAMOLECULAR BINDING OF N-TERMINAL PEPTIDE TO LYSINE BINDING SITE OF ITS OWN KRINGLE DOMAIN.

In order to clarify the possible candidate of lysine residue in N-terminal peptide region for intramolecular binding, we synthesized several peptides and examined those characteristics. In the sequence of N-terminal peptide of Glu[1] to Lys[76], only Lys[50] has negatively charged carboxyl radical of glutamic acid nearby (Scheme 1), which region is previously suggested as a candidate for the intramolecular binding to keep a tight conformation by Wiman and Wallén [15]. We synthesized a peptide of Ala[44]-Lys[50] (peptide 1) of Glu-plg and analyzed its effect on Glu-plg activation. In addition, we synthesized another peptide which contains additional Glu[51] (Ala[44]-Glu[51]) (peptide 2) in order to eliminate the possible function of C-terminal lysine of Ala[44]-Lys[50], and analyzed its function. Ala[44]-Ser[49] (peptide 3) was also synthesized to see the effect of Lys[50].

1. Effects of N-terminal peptides on the activation of Glu-plg by uPA.

The effects of synthesized peptides on the activation of Glu-plg by uPA were analyzed. The activation of Glu-plg by uPA was enhanced by peptide 1 and peptide 2 in a dose dependent manner (Fig. 1). The larger extent of the enhancement was obtained by peptide 1 than peptide 2. Peptide 3 did not show any enhancement (data not shown) suggesting that the presence of lysyl residue (Lys[50]) is required for the enhancement of the activation of Glu-plg by uPA. The effects of these peptides on the activation of Lys-plg is shown in Table 1. No enhancement was observed by these peptides and slight inhibition was observed. Slight inhibition in the presence of these peptides may have been caused by the competitive inhibition to urokinase or plasmin as EACA did. The fact that both peptide 1 and peptide 2 enhanced the activation of Glu-plg but did not enhance the activation of Lys-plg suggests that these peptide worked on Glu-plg molecule with a similar manner as lysine analogues by dissociating the intramolecular binding of N-terminal peptide.

Table 1. Effects of N-terminal peptides on Lys[77]-plasminogen activation

effectors (2mM)	activation rate 1×10^{-12} M/L/S
control	9.81 ± 0.45
peptide 1	6.53 ± 0.14
peptide 2	5.53 ± 0.34
EACA	7.45 ± 0.12

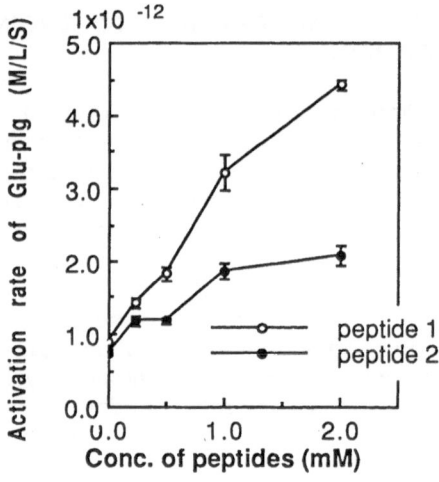

Glu-Plg (0.5 μM), S-2251 (0.3 mM) and the required concentration of the peptide was mixed in a buffer containing 50 mM Tris-HCl pH 7.4 and 100 mM NaCl in a final volume of 0.8 ml. The reaction was accelerated by the addition of 0.05 nM of uPA. Absorbances of *p*-nitroanilide were monitored at 405 nm, and were converted to the concentration of substrate hydrolyzed by employing an ε (1%, 1 cm, 405 nm) of 10,000 [16]. Data are shown by Mean ± SE (N=3).

FIG. 2. Effects of peptide 1 and 2 on the activation of Glu-plg by uPA.

2. Effects of N-terminal peptides on the conversion of Glu-plg to Lys-plg by plasmin.

Plasmin cleaves Lys^{76}-Lys^{77} peptide bond producing modified form of plasminogen, Lys^{77}-plg. The conversion of Glu-plg to Lys-plg by plasmin is enhanced by lysine analogues and the presence of fibrin [17]. We then analyzed the effects of synthesized N-terminal peptides on this conversion. The effects of both peptide 1 and peptide 2 on the conversion of Glu-plg form I by plasmin is shown in Figure 3. Both peptides enhanced the conversion as a function of the concentrations of peptides. The conversion of Glu-plg form II was also enhanced by the addition of both peptide 1 and peptide 2 (Data not shown), although the magnitude of the enhancement was smaller than that of Glu-plg form I. Peptide 3 did not show any enhancement on the conversion of both form I and form II of Glu-plg by plasmin (Data not shown). These data also consistent to the hypothesis that N-terminal peptide is binding to one of the lysine binding site of own kringle domain through $lysine^{50}$ residue, since it is proved that the dissociation of the intramolecular binding by either lysine analogues or fibrin amplifies this conversion [17].

N-terminal peptide region is considered to bind to lysine binding site of kringle 1 or kringle 4, or aminohexyl site in kringle 5 domain. Aminohexyl site is first proposed by Christensen as a candidate for the intramolecular binding [13]. She insisted that aminohexyl site can bind to positively charged side chain of amino acid mainly lysine residue even when it resides

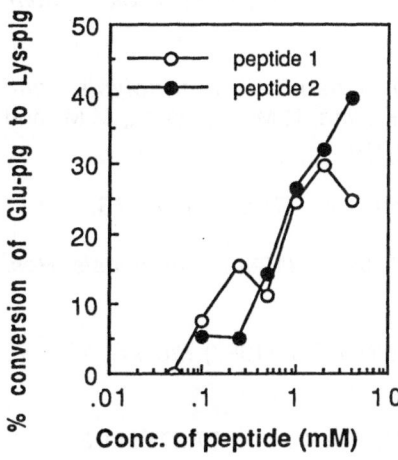

Conversion of Glu-plg to Lys-plg by plasmin was monitored by 7.5 % SDS-PAGE. Glu-plg (2.0 µM) was hydrolyzed by 1.9 nM of plasmin in the presence of desired concentrations of the peptides at 37 °C in a buffer containing 50 mM Tris-HCl pH 7.4 and 100 mM NaCl. The reaction was stopped by the addition of a sample buffer for SDS-PAGE and 10 µl of each sample was applied to SDS-PAGE gel. Stained gels by coomassie blue were subjected to the analysis by densitometer. Percent conversion of Glu-Plg to Lys-plg was calculated by the ratio of the density of Lys-plg to the sum of the densities of Glu-Plg and Lys-plg.

Fig. 3. Effects of peptide 1 and peptide 2 on the conversion of Glu-plg form I to Lys-plg form I by plasmin.

in the middle of the protein and that it prefers ligands which does not have free carboxyl function. Our data, however, showed that not only peptide 2 which carries lysine residue in the middle of the peptide but also peptide 1 which carries C-terminal lysine enhanced the activation of Glu-plg and the conversion of Glu-plg to Lys-plg with a similar manner, which suggests that lysine binding site is more likely the binding site than aminohexyl site for the N-terminal peptide region of Glu-plg.

In the present study we confirmed the hypothesis that intramolecular binding of N-terminal peptide to the lysine binding site of the kringle domain of own molecule keeping its tight conformation in the fluid phase. We also elucidated that lysine[50] of the N-terminal peptide region is the most probable candidate for this intramolecular binding.

REFERENCES

1. ROBBINS, K.C., SUMMARIA, L., HSIEH, B. and SHAH, R.J. (1967) *J. Biol. Chem.* *242*, 2333-2342.

2. CLAEYS, H., and VERMYLEN, J. (1974) *Biochim. Biophys. Acta* *342*, 351-359.

3.   VIOLAND, B.N., BYRNE, R., and CASTELLINO, F.J. (1978) *J. Biol. Chem.* *253*, 5395-5401.

4.   WALLÉN, P. (1978) In *Progress in Chemical Fibrinolysis and Thrombolysis, Vol 3,* (eds) Davidson, J. F.. Rowan, R.M., Samama, M.M. and Desnoyers, P.C., Raven Press. New York. p 167.

5.   TAKADA, A. and TAKADA, Y. (1980) *Thrombosis Res.* *18,* 167-176.

6.   TAKADA, A., TAKADA, Y. and SUGAWARA, Y. (1984) *Thrombosis Res.* *33,* 461-469.

7.   URANO, T., CHIBBER, B.A.K. and CASTELLINO, F.J. (1987) *Proc. Natl. Acad. Sci. U.S.A.* *84,* 4031-4036.

8.   URANO, T., de SERRANO, V.S., CHIBBER, B.A.K., and CASTELLINO, F.J. (1987) *J. Biol. Chem.* *262,* 15959-15964.

9.   URANO, T., de SERRANO, V. S., GAFFNEY, P. J. and CASTELLINO, F.J. (1988) *Biochemistry* *27,* 6522-6528.

10.   WIMAN, B. and COLLEN, D. (1978) *Nature* *272,* 548-549.

11.   LERCH, P. G., RICKLI, E. E., LERGIER, W. and GILLESSEN, D. (1980) *Eur. J. Biochem.* *107,* 7-13.

12.   CUMMINGS, H.S. and CASTELLINO, F.J. (1985) *Arch Biochem. Biophys.* *236,* 612-618.

13.   CHRISTENSEN, U. (1984) *Biochem. J.* *223,* 413-421.

14.   SJÖHOLM, I., WIMAN, B. and WALLÉN, P. (1973) *Eur. J. Biochem.* *39,* 471-479.

15.   WIMAN, B. and WALLÉN, P. (1975) *Eur. J. Biochem.* *50,* 489-494.

16.   CHIBBER, B.A.K., MORRIS, J.P. and CASTELLINO, F.J. (1985) *Biochemistry* *24,* 3429-3434.

17.   TAKADA, A. and TAKADA, Y. (1985) *Thrombosis Res.* *40,* 171-179.

INACTIVATION OF HUMAN TUMOR CELL PRO-UROKINASE BY GRANULOCYTE ELASTASE

N. KANAYAMA, T. TERAO[1], M. Schmitt, H. Graeff[2]

[1] Department of Obstetrics and Gynecology, Hamamatsu University School of Medicine, 3600 Handa-cho, Hamamatsu, 431-31, Japan
[2] Frauenklinik der Technischen Üniversitat, Ismaninger str. 22 München, FRG

KEYWORDS granulocyte elastase, pro-urokinase type plasminogen activator, fibrinolysis, uterine carcinoma, ovarian carcinoma

INTRODUCTION

Plasminogen activators are major mediators of pericellular proteolysis. In tissues of breast, prostate, cervix, and colon cancer, urokinase-type plasminogen activator (uPA) are produced and secreted as an enzymatically inactive single-chain proenzyme form (pro-uPA)[1-2]. The increase of uPA in tumor cells has been observed associated with increased tumor growth and metastatic potential[3]. Pro-uPA may be converted by small amounts of plasmin or cathepsin B into the enzymatically active two-chain form uPA (HMW-uPA) which subsequently converts plasminogen into the broad-spectrum serine protease plasmin. Plasmin degrades the fibrin-fibronectin matrix of the tumor stroma thus releasing fibrin remnants and cross-linked fibrin-fibronectin compounds[4].

In solution the naturally occuring plasminogen activator inhibitors PAI-1 and PAI-2 may inactivate uPA, but do not bind to pro-uPA[5-6]. Here we report that elastase released by chemotactically activated human granulocytes inactivates and degrades tumor cell pro-uPA.

MATERIALS AND METHODS

Reagents were purchased from the sources indicated in parenthesis. Pro-uPA from kidney tumor cell line TCL 598 (specific activity 135,000 units / mg) (SANDOZ, Nurnberg, FRG); purified human granulocyte elastase (Protogen, Laufelfingen, Switzerland); mouse monoclonal antibody to human granulocyte elastase (clone M752).

Sodium Dodecylsulphate Polyacrylamide Gel Electrophoresis (SDS-PAGE)

5 μg pro-uPA were incubated with 20 μl of 50 mM Tris-HCl, pH 7.3, containing elastase, thrombin, plasmin and trypsin (final concentration 500 nM) for 30 min at 37 ℃. The reaction mixture was subjected to SDS-PAGE according to Laemmli using 15 % polyacrylamide gels. Samples were heated (5 min, 100℃) in the presence of 10 mM 2-mercaptoethanol. Proteins were stained with Coomassie Blue R-250 and destained with 7 % acetic acid.

uPA-activity

To determine the effect of elastase on pro-uPA, 50 units of pro-uPA / 50 μl of 50 mM Tris-HCl, pH 7.3, was incubated with various concentrations of elastase.(0.5 - 200 nM) for 30 min at 37℃. The mixture will be referred as elastase treated pro-uPA samples. To convert the pro-uPA preparations to uPA, the samples were subsequently incubated with 2.8 nM plasmin (45 min, 37 ℃). Plasmin action was stopped by the addition of 200 KIU / ml aprotinin.

Samples containing elastase (0-200 nM) in Tris-HCl (in the absence of pro-uPA) added by plasmin served as control.

The enzymatic activity of uPA was measured with 0.33 mM of the chromogenic substrate S-2444. Enzyme activity was measured by the change in absorbance at 405 nM. The change of absorbance, produced by 50 units pro-uPA / 50 μl Tris-HCl which has been treated with 2.8 nM plasmin (45 min, 37 °C), was defined as 100 % HMW-uPA activity.

Fibrin-clot lysis assay was performed essentially to Kruithof et al[7]. Elastase treated pro-uPA samples were placed on top of the fibrin gels in the tube. The tubes were incubated (2 h, 37 °C). Lysis volume was calculated by the height of the dissolved gel. Lysis volume caused by 50 units pro-uPA / 50 μl Tris-HCl was defined as 100 % of fibrinolytic activity.

Stimulation of human peripheral blood granulocytes by N-formyl chemotactic peptide CHO-NLPNTL

Granulocytes were isolated essentialy as described[8]. Briefly, 50 ml of fresh human blood obtained from four healthy donors were anticoagulated by 5,000 units of heparin and immediately centrifuged through Ficoll-Hypaque, the cell pellet resuspended and residual erythrocytes lysed by 0.16 M NH_4Cl containing 12 mM $NaHCO_3$, 0.1 mM EDTA, pH 7.3. Granulocytes were washed with PBS. 5×10^6 granulocytes in 1 ml of phosphate buffered saline (PBS) containing 0.1 % BSA, 1 mg glucose, 1 mM $CaCl_2$, 0.5 mM $MgCl_2$, pH 7.3, were pretreated with cytochalasin B (5 μl / ml) for 5 min at 37°C. 10^{-8} M FNLPNTL was added and the cells incubated for 30 min at 37 °C. After low speed centrifugation (250 x g, 5 min), the supernatants were collected and then stored at -20 °C until measurement.

Immunohistochemical detection of uPA and elastase

Dewaxed paraffin-embedded tissue sections of uterine squemous and smear of ascites of ovarian carcinoma were processed for the detection of uPA and elastase. The slides were rinced in 50 mM Tris-HCl containing 125 mM NaCl, pH 8.1 (TBS) and were covered with 20 % rabbit serum (20 min 23 °C). Subsequently, a 1:50 dilution of monoclonal antibody to uPA (#394) or elastase was added in TBS. After washing in TBS the sections were reacted with a 1:50 dilution of Ig rabbit anti mouse Ig in TBS (30 min, 23 °C), washed with TBS and then a 1:50 dilution of mouse-APAAP in TBS was added (30 min, 23 °C). After washing in TBS the alkaline phosphatase-dependent staining was developed by 0.2 mg/ml Naphthol-AS-MX phosphate in combination with 10 mg/ml Fast Red TR in 0.2 M Tris-HCl (pH 8.5) containing 1 mM levamisole to block intrinsic alkaline phosphatase activity (20 min, 23 °C). The slide glasses were washed in TBS and water and mounted in glycerol-gelatine. Controls were performed by omitting the first antibody or by substituting the first antibody by irrelevant IgG-antibodies of the relevant species.

RESULTS
Degradation of pro-uPA by purified human granulocyte elastase

To test the effect of elaslase on tumor pro-uPA in vitro, elastase was incubated with purified pro-uPA and then the samples subjected to SDS-PAGE. Elastase degraded pro-uPA(M_r =55,000) into a molecule consisting of two major polypeotide chains of M_r =33,000 and M_r =22,000 connected by disulfide bond(s) similar to thrombin, plasmin and trypsin(Fig. 1). In no-reduced condition, this molecule produced by elastase treatment is a single band of M_r =55,000 (data not shown). Two bands of A chains were observed. Lower band of A chain is thougt to be a molecule which lacks EGF domain of pro-uPA. EGF domain of pro-uPA is split easily.

Efffect of proteases on pro-uPA activation

The elastase-treated samples were then subjected to plasmin treatment and tested for amidolytic or fibrinolytic activity. The new finding is that elastase treatment of pro-uPA also inhibited the conversion of pro-uPA by the subsequent addition of plasmin. Elastase has been discovered to digest plasminogen into smaller fragment(miniplasminogen). The fragment contains active site of plasmin and still has properties of plasmin especially amidolytic and fibrinolytic activities[14]. This makes ones able to assume that elastase has no effects on active site of plasmin. Therefore, inhibition of the conversion of pro-uPA to active uPA should be the direct effect of elastase on pro-uPA. The ability of pro-uPA to be converted by plasmin into an enzymatically active uPA-molecule (latent activity) decreased with increasing elastase concentration prior to plasmin treatment (Fig. 2). When pro-uPA was treated with elastase, first, and then plasmin added, the enzymatic activity of pro-uPA.was completely inhibited at 100 nM of elastase. Even at a concentration as low as 0.5 nM elastase, a loss of 35 % in activity was obsserved. When pro-uPA was incubated with plasmin, first, and then elastase (0 - 200 nM) added, no loss of pro-uPA activity was detected (data not shown). Samples containing elastase and plasmin did not dissolve fibrin clot.

Identification of elastase as the functional protease in granulocyte supernatants

When human peripheral blood granulocytes were stimulated in suspension with 10^{-8} M of the chemotactic peptide CHO-NLPNTL significant amounts of enzymatically active elastase was released. The addition of pro-uPA to supernatants obtained from stimulated granulocytes prevented the conversion of pro-uPA into enzymatically active uPA compare to supernatants obtained from unstimulated cells (Fig. 3). Supernatants of CHO-NLPNTL-stimulated granulocytes were incubated either with different concentrations of anti-elastase moAB (blocks activity) [0-15 μ M] . Fig. 4 shows the effect of cell supernatants obtained from stimulated granulocytes on amidolytic activity of pro-uPA in the presence of various amounts of moAB to elastase. With 15 μ M of moAB to elastase 80 % of the latent enzymatic activity was retained.

Localization of elastase and uPA in cancer tissues

Cells containing uPA or elastase were localized in human uterine cervical and asites from ovarian carcinoma by moAB to the antigens. uPA was localized in the cytoplasm of the cancer cells in squamous cell carcinoma and ascites from ovarian carcinoma (Fig. 5A 5B). Cancer cells stained homogeneously with the uPA antibody, although differences in staining intensity within the tumors were observed.

Elastase containing cells were localized in the tumor tissue and also the tumor stroma surrounding the tumor nests (Fig. 5C,5D). These cells most probably represent granulocytes. The staining pattern of some of these phagocytic cells was irregular indicating release of elastase into the tumor stroma and tumor cells.

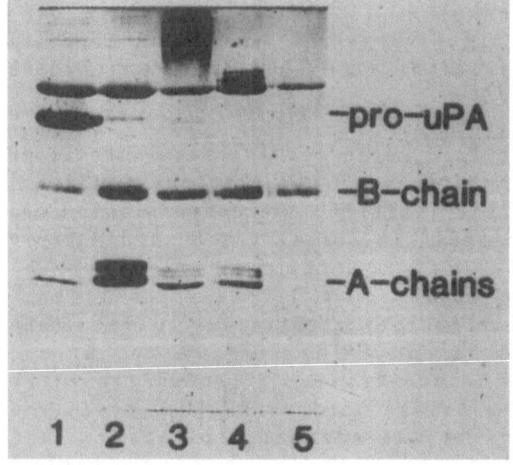

Fig.1 SDS PAGE analysis (15% acrylamide, reducing conditions) of pro uPA treated with proteases. 1:pro uPA, 2:pro uPA + elastase. 3:pro uPA + thrombin. 4:pro uPA + plasmin. 5:pro uPA + trypsin.

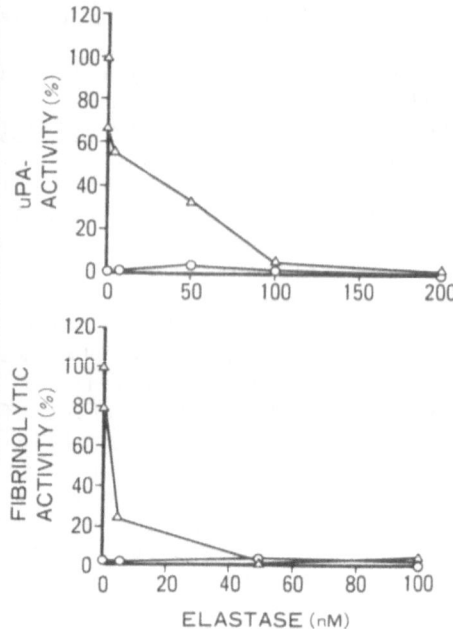

Fig.2 Effect of elastase on latent amidolytic and fibrinolytic activity of pro-uPA. △: pro-uPA(50 units) treated with various concentration of elastase. ○: elastase only(control).

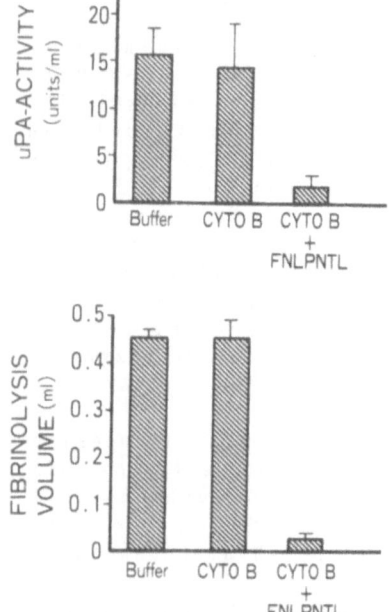

Fig.3 Effect of granulocyte supernatant on pro uPA. Supernatants from granulocytes treated with cytochalasin B(subactivated) or cytochalasin B + FNLPNTL(activated) were mixed with 100 units of pro uPA.

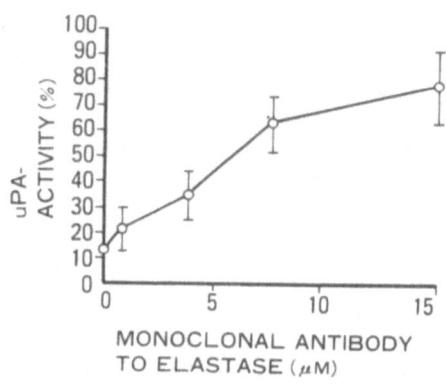

Fig.4 Inhibition of elastase activity in granulocyte supernatant by moAB to elastase.

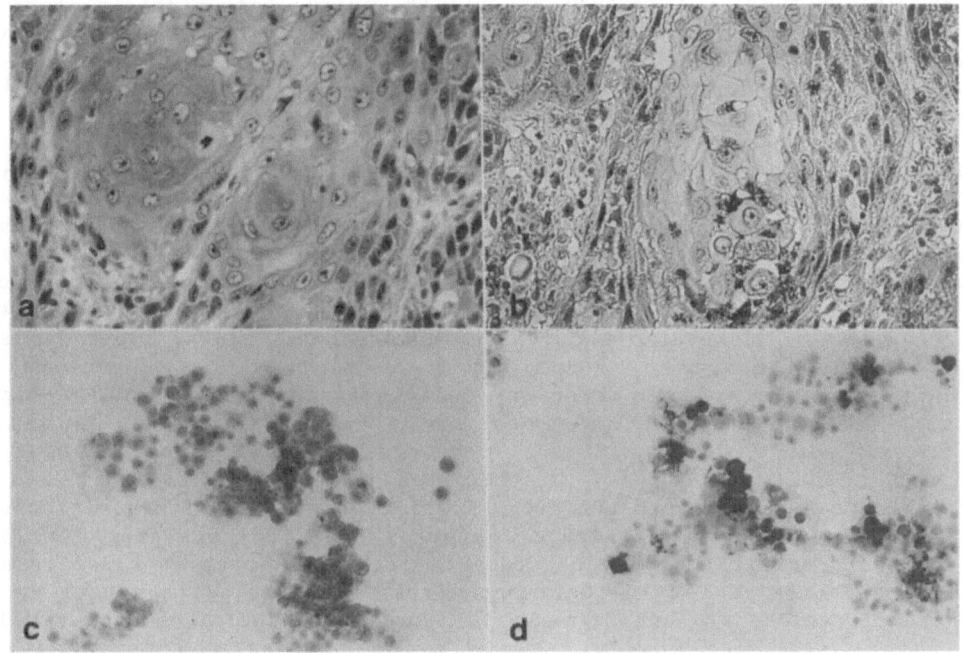

Fig. 5 Localization of elastase and uPA in uterine cancer and ascites from ovarian cancer. A:uPA in uterine cancer. B:elastase in uterine cancer. C:uPA in ascites from ovarian cancer. D:elastase in ascites from ovarian cancer. APPAP staining was performed.

DISCUSSION

Proteins in plasma and tissues such as fibrin(ogen), fibronectin, elastin and collagen, have been shown to be substrates for elastase. In this report we demonstrate that cancer tissues contain granulocytes which stain for elastase and tumor cells which stain for uPA. In in vitro experiments we have shown that purified tumor cell pro-uPA is degraded and inactivated by granulocyte elastase. Dose-dependent inactivation of pro-uPA by elastase was observed for amidolytic and fibrinolytic activity. However, inhibitory curve of elastase on uPA activity and fibrinolytic activity was different. This may be explained by the difference of active sites of the enzyme (uPA) in each reaction.

Treatment of pro-uPA with supernatants of chemotactically acivated granulocytes markedly depressed the latent enzymatic capacity of pro-uPA in vitro. By inhibition experiments with an antibody to the active site of elastase, elastase was identified as the key enzyme for this proteolytic activity. Heiple and Ossowski also reported on the destructive potential of cell supernatants of stimulated granulocytes on pro-uPA without identifying the proteases involved[9]. The inactivation capacity of the cell supernatants was inhibited by the serine protease inhibitor diisopropylfluorphosphate (DFP). Elastase is inactivated by

DFP. Heiple's and our results support the notion that elastase is one of the key enzymes in inactivating pro-uPA released by the cells. In vivo inactivation of pro-uPA by elastase most probably occurs in the extravascular space, because granulocyte is usualy activated after penetration of vessel.

Cancer tissue is a case where phagocytic cells such as granulocytes and pro-uPA containing tumor cells coexist[10]. Secreted pro-uPA is converted to enzymatically active HMW-uPA by small amounts of plasmin and then tumor-associated fibrinolysis can occur. Degradation products of the tumor stroma may attract phagocytic cells into tumor stroma. Recent studies demonstrated that even uPA itself is a potent chemotactic factor for granulocytes in vivo in addition to tumor stroma degradation products[11-12]. However, functions of granulocytes around tumor cells is little known compared with macrophage and lymphocytes. One of the functions attributed to granulocytes accumulated around tumor cells is the cytotoxic effect on tumor by releasing reactive oxidative intermediates upon stimulation. These oxidants are supposed to destroy tumor cells directly[13-14]. Now, we have found that activated granulocytes depress pro-uPA activity of cancer cells. Immunohistochemic al study support this finding. Immunohistochemical staining of cancer sections demonstrated that elastase is localized in cells close to uPA-rich cancer cells and in the tumor stroma surrounding the tumor nests. In those regions pro-uPA released by tumor cells could be inactivated by elastase set free from granulocytes. Dvorak et al. observed that granulocytes can attach to tumor cells and then elastase containing granules may be released[15]. The concentration of protease inhibitors in such regions should be lower than in plasma and would not be sufficient to neutralize elastase activity. Inactivation of pro-uPA by elastase may prevent the generation of enzymatically active uPA and thus diminish tumor cell metastasis and invasion.

REFERENCES
1. Corti A. Nolli ML. Soffientini A. Cassani G (1986) Trombo Haemastasis 56: 219-224
2. Stumo DC. Lijnen HR. Collen D (1986) J Biol Chem 261: 1274-1278
3. Markus G (1988) Enzyme 40: 158-172
4. Dvorak HF (1986) N Engl J Med 315: 1650-1659
5. Kruithof EKO. Tran-Thang C. Ransijin A. Bachmann F(1984)64:907-913
6. Kruithof EKO. Vassalli JD. Schleunig WD. Mattaliano RJ. Bachmann F (1986) J Biol Chem 261: 11207-11213
7. Kruithof EKO. Ransijin A. Bachmann F (1982) Thromb Res 28: 251-260
8. Borregaard N. Heiple JM. Simons ER. Clark RA (1983) J Cell Biol 97: 52-59
9. Heiple JM. Ossowski L (1986) J Exp Med 164: 826-840
10.Goldleski JJ. Lee RE. Leighton J (1970) Cancer Res 30: 1986-1993
11.Boyle MDP. Chdiodo VA. Lawman JP. Gee AR Young M (1987) 139: 169-174
12.Gudewicz PW. Gilboa N(1987)Biochem Biophys Res Commun 147:1175-1181
13.Gerrad TL. Cohen DJ. Kaplan AM(1981)J Natl Cancer Inst 66: 483-488
14.Clark RA. Klebanoff SJ J Exp Med 141: 1442-1447
15.Dvorak AM. Connell AB. Proppe K. Dvorak HF (1978) J Immunol 120:1240-1248

CATHEPSIN B EFFICIENTLY ACTIVETS THE SOLUBLE AND THE TUMOR CELL RECEPTOR-BOUND FORM OF THE PROENZYME UROKINASE-TYPE PLASMINOGEN ACTIVATOR (PRO-UPA)

H. KOBAYASHI [1], N. KANAYAMA [1], M. SCHMITT [2], L. GORETZKI [2], N. CHUCHOLOWSKI [2], J. CALVETE [3], M. KRAMER [4], W.A. GÜNZLER [5], F. JÄNICKE [2], T. TERAO [1], H. GRAEFF [2].

1 Department of Obstetrics and Gynecology, Hamamatsu University School of Medicine, Hamamatsu, Shizuoka, 431-31, Japan, 2 Frauenklinik der Technischen Universität München im Klinikum rechts der Isar, D-8000 München 80, the 3 Max-Plank-Institut für Biochemie, D-8033 Martinsried bei München, the 4 Dermatologisches Institut der Universität Heidelberg, D-6900 Heidelberg, and 5 Grünenthal GmbH, D-5190 Stolberg, FRG.

INTRODUCTION

Tumor cell invasion and metastasis is a multifactorial process, which at each step may require the action of proteolytic enzymes such as collagenase, cathepsins, plasmin, or plasminogen activators [1,2]. An enzymatically inactive proenzyme form of the urokinase-type plasminogen activator (pro-uPA) is secreted by tumor cells [1,3]. Cathepsin B, a cysteine-dependent protease, which is elevated in tumors, plays a regulatory role in collagen degradation, since it can convert inactive procollagenase IV to its enzymatically active form [4]. We demonstrate that cathepsin B has the capacity to efficiently convert soluble or tumor cell receptor-bound pro-uPA to enzymatically active two-chain uPA. Thus, the cellular protease cathepsin B may substitute for the plasma protease plasmin in the activation of pro-uPA released by tumor cells.

MATERIALS AND METHODS

Enzymatic activity   The chromogenic substrate S-2444 was applied to determine uPA activity. The uPA activities of the samples were calculated in Ploug units. Interaction of pro-uPA with cathepsin B   10 μM pro-uPA was incubated with increasing concentrations of cathepsin B (0-20 μM, 40 min, 37 $^\circ C$). Activation of pro-uPA by cathepsin B in the presence of cysteine or the inhibitor E-64 was examined by SDS-PAGE, enzymatic assays, and reversed-phase HPLC.
Plasminogen activator casein plaque assay   A proteolytic plaque assay was performed. The plaques detected around the holes represent zone of casein lysis [5]. SDS-PAGE and Western blot   Pro-uPA was incubated with cathepsin B, cathepsin D, or plasmin. The reaction mixture was analyzed by SDS-PAGE under the reducing conditions [6]. Proteins separated by SDS-PAGE were electroblotted onto polyvinylidine difluoride membrane. Monoclonal antibody (moAB) 377 (specific for an epitope within the A-chain of uPA), moAB GFD1 (specific for a peptide sequence within

115

the growth factor-like domain of uPA), and moAB 98.6 (specific for an epitope
within the B-chain of uPA) were used for Western blot.

Isolation of A- and B-chain of cathepsin B-treated pro-uPA   Isolation was performed by reversed-phase HPLC, and the cleavage site was determined by N-terminal
amino acid sequence analysis.

Cell culture   Promyeloid U937 cells were grown in RPMI 1640 medium supplemented
with 10 % fetal calf serum. Differentiated U937 cells were obtained by incubation in the above medium containing 1 μM phorbol 12-myristate-13-acetate (PMA).

Binding of cathepsin B-treated pro-uPA to the uPA receptor on PMA-stimulated
U937 cells were assessed by Flow cytometry. 1.5 nM FITC conjugated pro-uPA was
applied to the resuspended PMA-stimulated U937 cells. Cell-associated fluorescence was measured (FACScan).

Cathepsin B-mediated cleavage of receptor-bound pro-uPA   PMA-stimulated U937
cells were incubated with pro-uPA, washed and then incubated with cathepsin B
(0-0.5 μg/ml). Cell-bound uPA was obtained by acid treatment and then concentrated. Samples were applied to SDS-PAGE and analyzed by Western blot.

RESULTS

    ACTIVATION OF PRO-UPA BY CATHEPSIN B   Proteases such as elastase, cathepsin B,
plasmin, thrombin, collagenase, and trypsin convert pro-uPA to two-chain uPA
(Fig. 1). Plasmin treatment renders to enzymatically active high molecular weight-
uPA (HMW-uPA); elastase [7] and thrombin [8] treatment renders an inactive two-chain
uPA-molecules. Cathepsin B and D are tumor-associated proteases also produced
and released by tumor cells [9]. To investigate the proteolytic action of cathens-
in B on pro-uPA, recombinant pro-uPA was incubated with different concentrations
of cathepsin B or D, and as a control with plasmin, respectively. Cathepsin B
was found to be a good activator of pro-uPA, similar to plasmin (Fig. 2). Preincubation of cathepsin B with the inhibitor E-64 prevented the cathepsin B-induc-
ed activation of pro-uPA. Also, the action depends on cysteine in the buffer.
Cathepsin B action is not affected in the presence of Trasyrol. When HMW-uPA was
further incubated with cathepsin B, no loss of activity was seen. We also analyzed the cathepsin B-generated uPA activity toward the protein casein, indicating
that pro-uPA activated by cathepsin B exhibits both amidolytic and proteolytic
activity.

    MOLECULAR CHARACTERIZATION AND IDENTIFICATION OF THE CATHEPSIN B-MEDIATED CLE-
AVAGE SITE IN PRO-UPA   Proteolysis of pro-uPA with cathepsin B was visualized
by SDS-PAGE and Western blot. Virtually complete proteolysis was obtained within
160 min, yielding two major polypeptides of MW=33,000 (B-cahin) and MW=22,000
(A-chain). The results of Western blot analyses are illustrated in Fig. 3, indicating that MW=33,000 and MW=22,000 correspond to the B-chain and A-chain of uPA,

117

respectively. The A- and B-chain of cathepsin B-treated pro-uPA were separated
by reversed-phase HPLC and analyzed for their N-terminal amino acid sequences.
The cleavage site was identified at position Lys^{158}-Ile^{159}, which is identical
with the cleavage site for plasmin.

CATHEPSIN B ACTIVATES SOLUBLE AS WELL AS TUMOR CELL RECEPTOR-BOUND PRO-UPA

The binding of pro-uPA or cathepsin B-cleaved pro-uPA to the surface receptor on
promyeloid U937 cells has been compared in a competition assay utilizing FCM,
indicating that cathepsin B-cleaved pro-uPA binding to the uPA receptor does not
significantly differ from that of pro-uPA. In addition, receptor-bound pro-uPA
is sensitive to cathepsin B treatment. Receptor-bound pro-uPA was completely
converted into two-chain uPA at a cathepsin B concentration of 0.5 µg/ml, suggesting that soluble and receptor bound pro-uPA can be activated by exogenous
cathepsin B (Fig. 4).

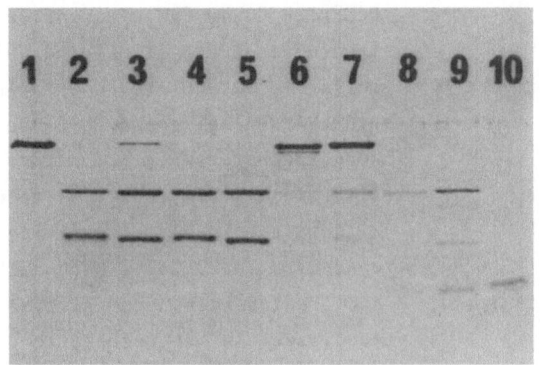

Fig. 1   CLEAVAGE OF RECOMBINANT
PRO-UPA BY PROTEASES. Pro-uPA was
incubated with elastase (lane 2),
cathepsin B (lane 3), plasmin
(lane 4), thrombin (lane 5), V8
protease (lane 6), collagenase
(lane 7), pronase (lane 8), trypsin
(lane 9), pepsin (lane 10), respectively, and analyzed by SDS-PAGE
under the reducing con-ditions.
Lane 1 represents the electrophoretic mobility of pro-uPA.

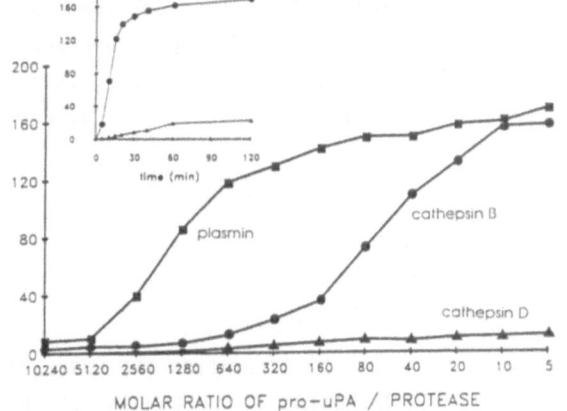

Fig. 2   DOSE-DEPENDENT ACTIVATION
OF PRO-UPA BY CATHEPSIN B OR D.
Pro-uPA was incubated with different concentration of cathepsin B
(●), cathepsin D (▲), or, as a
control, plasmin (■).

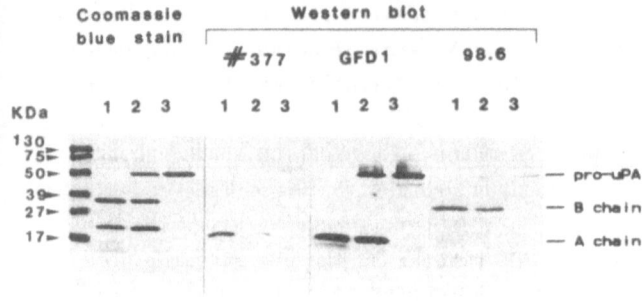

Fig. 3 COMPARISON BY WESTERN BLOT ANALYSIS OF UPA FRAGMENTS OBTAINED AFTER TREATMENT OF PRO-UPA BY CATHEPSIN B, D, OR PLASMIN.

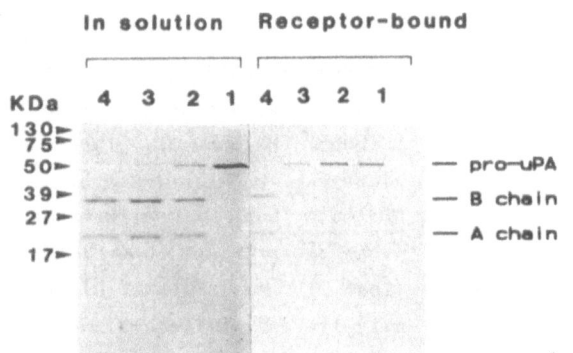

Fig. 4 CATHEPSIN B CLEAVAGE OF RECEPTOR-BOUND PRO-UPA. After acid treatment of PMA-stimulated U937 cells, to dissociate receptor-bound pro-uPA/uPA molecules, U937 cells were incubated with 7.4 nM pro-uPA. After washes, cells were treated with cathepsin B at increasing concentrations. Cells were washed and treated with 50 mM glycine-HCl to dissociate the receptor-bound uPA molecules. Samples were analyzed by SDS-PAGE under reducing conditions. Cathepsin B concentration (μg/ml): lane 1, 0; lane 2, 0.05; lane 3, 0.1; lane 4, 0.5.

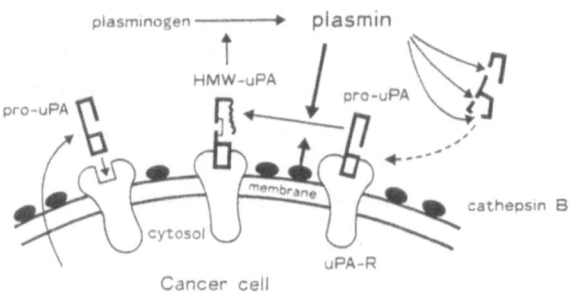

Fig. 5 ACTIVATION AND REGULATION MECHANISM OF UPA BINDING TO THE UPA RECEPTOR.

DISCUSSION

The major finding is that enzymatically inactive pro-uPA can be cleaved and activated by cathepsin B. Similar to plasmin, pro-uPA is converted by cathepsin B into two-chain HMW-uPA with equal efficiency in solution or bound to receptors on the tumor cell surface. Plasmin also kallikrein are known to cleave pro-uPA at amino acid position Lys[158]-Ile[159] [7,9]. The same peptide bond is cleaved by cathepsin B during the activation of pro-uPA. In contrast, cathepsin D was ineffective. The Lys[158]-Ile[159] bond is a common activation site of serine proteases. The action of cathepsin B on pro-uPA depends on cysteine and is inhibited by E-64. Trasyrol, an inhibitor for plasmin, had no effect.

Previous biochemical studies of the subcellular distribution of cathepsin B in tumor cells indicated the presence of enzymatically active cathepsin B in the plasma membrane of cancer cells, suggesting that cathepsin B is a membrane-associated protein in tumor cells [10]. The extracellular cell-associated cathepsin B can activate latent tissue collagenase IV (procollagenase IV), resulting in the degradation of the tumor stroma and the basement membrane [4]. Moreover, the proteolytic conversion of procathepsin B to the enzymatically active cathepsin B was demoinstrated to be due to cathepsin D [11]. Cathepsin D also exerts an important function in the increased activity of cysteine protease through inactivation of cystatin, a natural inhibitor of cathepsin B [11]. It could be shown by several clinical studies that determination of cathepsin D antigen may serve as an independent prognostic factor [12].

Plasminogen activators that convert inactive plasminogen into the serine protease plasmin are also involved in tumor stroma degradation and tumor cell metastasis. uPA is secreted by tumor cells and binds to cell surface receptors on tumor cells in an autocrine fashion [1,3]. The cell surface-focused uPA system converts plasminogen to plasmin, which also binds receptors on tumor cells resulting in a cellular surface protease system facilitating the degradation of the surrounding tumor stromas and the basement membrane. In fact, elevated levels of pro-uPA/uPA have been determined in cancer of the breast, ovary, prostate, and the digestive tract [13]. The clinical relevance of these findings was demonstrated by Janicke et al. [14] and Duffy et al. [15], indicating the uPA antigen to be an independent prognostic factor in breast cancer. In these reports, it was shown that high pro-uPA/uPA levels are correlated with increased risk for early recurrence and shorter survival. So far, detailed studies on the interaction of cathepsin B, D, and pro-uPA/uPA levels in cancer patients and outcome of the disease have not been undertaken.

Evidently, independent mechanisms exhist which may lead to activation of pro-uPA on tumor cell surfaces and thus plasmin generation:

1 Intrinsic autocatalytic plasminogen-activating activity of pro-uPA, several

orders of magnitude lower than of two-chain HMW-uPA [16].

2 Activation of pro-uPA by plasma proteases (plasmin, kallikrein, and Factor XIIa) [1]; and

3 Activation of pro-uPA by the tumor cell-associated protease cathepsin B

REFERENCES

1. Dano K., Andreasen P.A., Grondahl-Hansen J., Kristensen P.I., Nielsen L.S. and Striver L. (1985) Adv. Cancer Res. 44: 139-226.

2. Liotta L.A., Rao C.N. and Barsky S.H. (1983) Lab. Invest. 49: 636-649.

3. Stoppelli M.P., Corti A., Soffientini A., Cassani G., Blasi F. and Assoian R.K. (1985) Proc. Natl. Acad. Sci. USA 82: 4939-4943.

4. Eeckhout Y. and Vaes G. (1977) Biochem. J. 166: 21-31.

5. Estreicher A., Wohlwend A., Belin D., Schleuning W.D. and Vassalli J.D. (1989) J. Biol. Chem. 264: 1180-1189.

6. Laemmli U.K. (1970) Nature 227: 680-685.

7. Schmitt M. Kanayama N., Henschen A., Hollrider A., Hafter R., Gulba D., Janicke F. and Graeff H. (1989) FEBS Lett. 255: 83-88.

8. Ichinose A., Fujikawa K. and Suyama T. (1986) J. Biol. Chem. 261: 3486-3489.

9. Gurewich V. and Pannell R. (1987) Blood 69: 969-972.

10. Krepela E., Bartek J., Skalkova D., Vicar J., Rasnick D., Taylor-Papadimitriou J. and Hallowes R.C. (1987) J. Cell Sci. 87: 145-154.

11. Nishimura Y. and Kato K. (1987) Biochem. Biophys. Res. Commun. 148: 254-259.

12. Tandon A.T., Clark G.M., Chamness G.C., Chirgwin J.M. and McGuire W.L. (1990) New Engl. J. Med. 322: 297-302.

13. Markus G. (1988) Enzyme (Basel) 40: 158-172.

14. Janicke F., Schmitt M., Ulm K., Gossner W. and Graeff H. (1989) Lancet 11: 1149.

15. Duffy M., O'Grady P., Devaney D., O'Siorain L., Fennelly J.J. and Lijnen H.R. (1988) Cancer 62: 531-533.

16. Lijnen H.R., Van Hoef B., Nelles L. and Collen D. (1990) J. Biol. Chem. 265: 5232-5236.

Circulation

HEMOSTATIC DISTURBANCES IN PATIENTS WITH OCCLUSIVE VASCULAR DISEASE

M.BIELAWIEC, M.Z.WOJTUKIEWICZ, K.KRUPIŃSKI, J.KŁOCZKO, A.BODZENTA-
ŁUKASZYK, A.SZPAK AND B.DWORAK
Department of Hematology, Medical School, Bialystok, Poland

Key words: atherosclerosis, coagulation, fibrinolysis, platelets,
            leukocytes

INTRODUCTION

Atherosclerosis, because of its high prevalence, creates an important
clinical problem. One of the most interesting and intriguing concepts
constructed to explain the pathophysiology of atherosclerosis is
"thrombogenic" or "encrustation" theory. Even though numerous reports
on the role of hemostatic mechanisms in the development of atherosclerosis have appeared since the time of Rokitansky the contribution of
hemostasis to atherosclerotic process is far from being fully understood (reviewed in 1). Platelet material and fibrin content in atherosclerotic plaques as well as large amounts of fibrin present in arterial thrombi have been reported (2 and references therein). Recently, leukocyte contribution to atherogenesis increasingly attracts
attention for leukocyte adhesion to the endothelium is one of the
first events to occur in experimental models of atherosclerosis (3).
In the present work we focused our interest on global evaluation of
clotting and fibrinolytic system as well as platelet, and leukocyte
functions in patients with peripheral occlusive arterial disease.

MATERIAL AND METHODS

The study was carried out on patients with peripheral occlusive arterial disease (POAD) of the lower limbs in 2nd and 3rd degree of the
disease according to Leriche-Fontaine. The diagnosis of the peripheral
occlusive arterial disease was made in all cases on the basis of medical examination, oscillography, impedance plethysmography, Doppler-US-
measurement and tissue oxygen pressure examination. The control group
comprised healthy subjects in the same age range.
Factor XIII activity was determined according to Lorand et al. (4).
Plasma fibronectin, antithrombin III (AT III), $alpha_2$-antiplasmin
($alpha_2$-AP), $alpha_2$-macroglobulin   ($alpha_2$-M), $alpha_1$-antitrypsin
($alpha_1$-AT), C_1-esterase inhibitor (C_1-I) and plasminogen concentra-
tions were measured by means of rocket immunoelectrophoresis using
commercial antisera (Behring).
Fibrinogen was assayed as a clottable protein by means of phenol rea-
gent (5).
Euglobulin lysis time was measured acc. to Chakrabarti et al. (6).
Serum FDP concentration was estimated acc. to Merskey et al. (7).
Platelet function tests:
Platelet adhesion to bovine subendothelial extracellular matrix (ECM)
was studied using ECM produced by IBT Int.Bio-Technologies Ltd. Kiryat
Hadassah Jerusalem, Israel (8). The plastic dishes with a diameter of
35 mm covered by bovine extracellular matrix were washed 3 times with
phosphate buffered saline (PBS before use. Them ECM were covered with
2 ml platelet rich plasma (PRP). The PRP were diluted with platelet
poor plasma (PPP) before investigation to obtain 200.000 - 250.000
platelets/µl. After 30 min incubation of the PRP with ECM, the PRP was
removed and the plastic dishes were carefully washed 20-times in NaCl
citrate solution (9:1). Then, platelets adhering to ECM were fixed by
5 min incubation in 6% glutardialdehyde. For staining the plastic dishes were put for 5 min into 0,1 N $KMnO_4$ and then, after 20 times washing with destilled water, to a Giemsa solution (1:5) for 30 min.

Finally, after 20-times washing in destilled water, the platelets adhering to ECM were counted by direct microscopic observation (occular 12,5-fold, objective 25 fold). Each PRP was tested on 5 matrix dishes. Triple counting was performed on each sample. The investigated area for the single counting was 0,137 mm^2 (total area 2,055 mm^2).
Platelet adhesion to the siliconized glass was studied according to the method described by Breddin (9).
Platelet spraeding was studied acc. to Breddin and Bürck (10).
Platelet aggregation was performed on Elvi-840 aggregometer, Elvi-Logos,Milan,Italy, using Born's method (11); PRP samples were tested for their responsiveness to appropriate treshold aggregating concentration of ADP (TAC of ADP).
Beta-thromboglobulin (beta-TG) and cyclic AMP were assayed with the use of commercially available radioimmunoassay kits (Amersham, England).
Thromboxane B_2 (TxB_2) and 6-keto-PGF$_{1\alpha}$ were measured by using radioimmunoassay kits (New England Nuclear, USA).
The total number of platelet-leukocyte aggregates was investigated according to the method of Silbergleit in our modification (12). The white blood cells (WBC) adherence to the nylon fibres was performed using Mac Gregor's method (13).
Statistical analysis included Student's t-test (comparisons between groups).All assay variables are represented by their mean and standard deviation.

RESULTS
The POAD patients revealed significantly elevated fibrinogen, factor XIII and fibronectin concentrations (Table 1). They had also markedly increased plasminogen level and significantly prolonged euglobulin lysis time while FDP concentration was not significantly different from that in the control group (Table 1).

Table 1. Coagulation and fibrinolytic parameters in patients with POAD in comparison with healthy controls.

	Fibrinogen g/L	Fibronectin %	F.XIII %	Plasminogen %	FDP µg/ml	Euglobulin. lysis time min
Control (n=35)	$3,6\pm0,8$	94 ± 20	98 ± 15	88 ± 16	$4,3\pm2,1$	228 ± 59
POAD (n=50)	$4,3\pm1,2$	111 ± 29	118 ± 27	111 ± 26	$5,0\pm3,8$	337 ± 204
Statistical significance	$p<0,01$	$p<0,01$	$p<0,001$	$p<0,001$	NS	$p<0,01$

The patients showed a significant increase in alpha$_1$-antitrypsin and C$_1$-esterase inhibitor concentrations, whereas there were no significant changes in antithrombin III (AT III), alpha$_2$-macroglobulin (alpha$_2$-M) and alpha$_2$-antiplasmin (alpha$_2$-AP) level (Table 2).
Patients suffering from peripheral occlusive arterial disease revealed increased platelet adhesion to the siliconized glass, subendothelial extracellular matrix and enhanced platelet spraeding as compared to the control group (Table 3). Platelet adhesion to the subendothelial extracellular matrix was significantly higher in POAD group in comparison with healthy subjects.
Platelet aggregation induced by ADP was higher in patients suffering from peripheral occlusive disease (Table 4).

Table 2. Serine protease inhibitors in patients with POAD in comparison with healthy controls.

	AT III %	α_2-M %	α_1-AT %	C_1-I %	α_2-AP %
Control (n=35)	97 ± 16	101 ± 24	100 ± 22	93 ± 20	99 ± 16
POAD (n=50)	101 ± 13	98 ± 29	127 ± 39	117 ± 27	103 ± 26
Statistical significance	NS	NS	$p<0,001$	$p<0,001$	NS

Table 3. Platelet adhesion to the siliconized glass, subendothelial extracellular matrix and platelet spraeding in patients with POAD and control group.

	Platelet adhesion to the siliconized glass Index	Platelet spraeding %o spraed forms	Platelet adhesion to the bovine ECM % inhibition compared to the control
Control (n=28)	$0,83\pm0,12$	$380,6\pm45,6$	0
POAD (n=36)	$1,52\pm0,25$	$660,5\pm52,3$	$+18,4\pm5,7$
Statistical significance	$p<0,001$	$p<0,01$	$p<0,01$

Table 4. Platelet aggregation, activity of stable metabolites of TxA_2 - TxB_2 and PGI_2 - 6-keto-$PGF_{1\alpha}$, cyclic AMP and beta-TG in patients with POAD in comparison with control group.

	Control (n=28)	POAD (n=72)	Statistical significance
Beta-TG ng/ml plasma	$47,5\pm15,4$	$78,5\pm20,8$	$p<0,05$
TAC of ADP µM	$1,24\pm0,34$	$0,38\pm0,15$	$p<0,001$
TxB_2 pg/ml plasma	$91,4\pm40,2$	$221,4\pm92,5$	$p<0,01$
6-keto-$PGF_{1\alpha}$ pg/ml plasma	$103,5\pm38,5$	$41,2\pm20,5$	$p<0,01$
cAMP pM/ml plasma	$51,5\pm5,6$	$34,2\pm11,8$	$p<0,01$

The patients showed increased concentrations of TxB$_2$ and beta-TG as well as a decrease in the level of plasma cyclic AMP and concentration of 6-keto-PGF$_{1\alpha}$ (Table 4).

The patients with POAD revealed increased mean numbers of platelet-leukocyte aggregates as compared to the control group, while the leukocyte adherence to the nylon fibres was not significantly enhanced (Table 5).

Table 5. Leukocyte adherence to the nylon fibres and total number of platelet and leukocyte aggregates in patients with POAD in comparison with control group.

	Leukocyte adherence to the nylon fibres %	Total number of platelet-leukocyte aggregates number of aggregates/µl
Control	$15,3 \pm 2,1$ (n=37)	$37,5 \pm 12,4$ (n=45)
POAD	$18,5 \pm 3,2$ (n=37)	$174,5 \pm 37,5$ (n=45)
Statistical significance	$p > 0,05$	$p < 0,001$

DISCUSSION

The observed changes in coagulation and fibrinolysis in atherosclerotic patients indicate a hypercoagulable tendency with inadequate fibrinolysis, what may promote intravascular and endoparietal fibrin deposition. Factor XIII strengthens fibrin against fibrinolytic degradation by formation of cross-links between fibrin molecules and incorporation of alpha$_2$-antiplasmin, main plasmin inhibitor, into the clot (14). Moreover, F.XIII catalyzes formation of cross-links between fibrinogen, fibronectin and collagen (15). Curiously enough, we found elevated F.XIII level concomitant with hyperfibrinogenemia. As hyperfibrinogenemia may reflect activation of blood coagulation one would rather expect a decrease in F.XIII level. It is possible that overproduction of F.XIII may account for this phenomenon. It may represent a lipoprotein-mediated mechanism (16) which may be an additional factor favouring development of atherosclerotic plaque.

We found several alterations of platelet function including enhanced concentration of platelet release products (beta-TG), increased levels of prostanoid pathway products (TxB$_2$), increased platelet adhesiveness and aggregability "in vitro". Observed changes may imply a pathogenetic significance of platelets in atherosclerotic process. A variety of platelet function abnormalities was reported by other authors (reviewed in 17,18). However, pathomechanism of changes in platelet behaviour is not known. Forconi et al. suggested a possibility of local platelet activation under ischemic conditions (19). Platelet dysfunction may as well result from shear stress at stenotically changed arteries or depend, to some extent, on higher hematocrite and/or elevated fibrinogen level. There is still a possibility of platelet activation by leukocytes (20).

Prostacyclin, a potent inhibitor of platelet aggregation and a strong vasodilator, is at present considered to be of prime importance for normal blood flow maintenance. Consequently, a decreased concentration of 6-keto-PGF$_{1\alpha}$, a stable prostacyclin metabolite, observed in POAD patients may indicate rather poor defence mechanism against vascular occlusion in this disease. Endothelial cell is responsible for

production of prostacyclin as well as tissue plasminogen activator
(t-PA) and its inhibitor (PAI-1). Therefore, a decreased 6-keto-PGF$_{1\alpha}$
concentration as well as prolonged euglobulin lysis time (indicative
of a decrease in t-PA and/or an increase in PAI-1 level) may be viewed as endothelial cell dysfunction. Other authors reported on defective prostacyclin production as an important factor in atherogensis
(21).

We found increased leukocyte activation "in vitro" (enhanced leuko-
cyte-platelet aggregates formation) in patients with POAD. It raises
a possibility of leukocytes implication in atherogensis by their increased adhesion to the endothelium and enhanced migration into ischemic tissues. Activated leukocytes are capable of producing and secre-
ting proteolytic enzymes. Thus, an increased concentration of alpha$_1$-
antitrypsin, main inhibitor of granulocytic proteases, found in POAD
patients may reflect a defence mechanism against neutrophilic proteases. Activated leukocytes may favour atherosclerotic process as well
as microthrombosis and ischemic changes in the microcirculation by
production of platelet activating factor, leukotrienes, superoxide
anions and proteolytic enzymes leading to platelet activation and dis-
ruption of endothelial cells with subsequent increase in vascular per-
meability, cytolysis and vasospasm (22).

Our results confirm contribution of hemostatic abnormalities to
the pathophysiology of atherosclerosis. However, further studies are
required to assess more precisely pathogenetic significance of hemostatic system in atherogenesis. Based on our data and reviewed lite-
rature it seems reasonable that correction of disturbed coagulation,
fibrinolysis, endothelial, platelet and leukocyte functions by means
of pharmacological intervention may be of value in the prophylaxis
of atherosclerosis and treatment of patients with peripheral occlusive
arterial disease.

REFERENCES

1. Ulutin O.N. (1986) Atherosclerosis and haemostasis. Semin.Thromb.
   Hemost. 12:156-174.

2. Bini A., Fenoglio J.Jr., Sobel J., Owen M. and Kaplan K.L. (1987)
   Immunochemical characterization of fibrinogen, fibrin I, and fibrin
   II in human thrombi and atherosclerotic lesions. Blood 69:1038-1045

3. Ross R. (1986) The pathogenesis of atherosclerosis. N.Engl.J.Med.
   314:488-500.

4. Lorand L., Urayama T., de Kiewiet J.W.C. and Nossel H.L.(1969)
   Diagnostic and genetic studies of fibrin-stabilizing factor with a
   new assay based on amine incorporation. J.Clin.Invest. 48:1054-1064

5. Folin O. and Ciocalteau V. (1927) Tyrosine and tryptophan determination in proteins. J.Biol.Chem. 73:627-650.

6. Chakrabarti R., Bielawiec M., Evans I.F. and Fearnley G.R.(1968)
   Methodological study and recommended technique for determining the
   euglobulin lysis time. J.Clin.Pathol. 21:698-701.

7. Merskey C., Lalezari P. and Johnson A.J.(1969) A rapid, simple,
   sensitive method for measuring of fibrinolytic split products in
   human serum. Proc.Soc.Exp.Biol.Med. 131:871-875.

8. Vlodavsky I., Eldor A., Hy Am E., Atumon R. and Fuchs Z. (1982)
   Platelet interaction with the extracellular matrix produced by cultured endothelial cells: A model to study the thrombogenicity of
   isolated subendothelial basal lamina. Thromb.Res. 28:179-185.

9. Breddin H.K. (1964) Zur Messung der Thrombozytenadhäsivität. Thrombos.Diathes.haemorrh. 12:269-281.

10. Breddin H.K., Bürck K.H. (1963) Zur Klinik der Thrombozyten funktionsstörung uter besonderer Berücksichtigung der Ausbreitungsfähigkeit der Thrombozyten an silikonisiertem Glass. Thrombos. Diathes.haemorrh. 9:525-545.

11. Born G.V.R. (1962) Aggregation of blood platelets by adenosine diphosphate and its reversal. Nature 194:927-931.

12. Bielawiec M., Kiersnowska-Rogowska B., Myśliwiec M. and Perzanowski A. (1975) The behaviour of the platelet-leukocyte aggregation in thrombotic conditions. Folia Haematol. 102:220-227.

13. MacGregor R.R., Macarek E.J. and Kefalides N.A. (1976) Comparative adherence of granulocytes to endothelial monolayers and to the nylon fiber. J.Clin.Invest. 61:697-704.

14. Sakata Y. and Aoki N. (1980) Cross-linking of alpha$_2$-plasmin inhibitor to fibrin by fibrin-stabilizing factor.J.Clin.Invest. 65:290-297.

15. Lorand L., Losowsky M.D. and Miloszewski K.J.M. (1980) Human factor XIII: Fibrin-Stabilizing Factor. Prog.Hemost.Thromb. 5:245-290

16. Cucuianu M.P., Miloszewski K., Porutiu D. and Losowsky M.S. (1976) Plasma factor XIII and platelet factor XIII in hyperlipaemia. Thromb.Haemost. 36:542-550.

17. Sinzinger H. (1986) Role of platelets in atherosclerosis. Semin. Thromb.Hemost. 12:124-133.

18. Hoak J.C. (1988) Platelets and atherosclerosis. Semin.Thromb. Hemost. 14:202-205.

19. Forconi S., Pieragalli D., Guerrini M. and DiPerri T. (1987) Hemorheology and peripheral arterial diseases. Clin.Hemorheol. 7:145-158.

20. Mehta P., Mehta J., Lawson D. and Krop I.(1986) Leukotrienes potentiate the effects of epinephrine and thrombin on human platelet aggregation. Thromb.Res. 41:731-738.

21. D'Angelo Y., Villa S., Myśliwiec M., Donati M.B. and de Gaetano G. (1978) Defective fibrinolytic and prostacyclin like activity in human atheromatous plaques. Thromb.Haemostas. 39:535-536.

22. Ernst E., Hammerschmidt D.E., Bagge U., Matrai A. and Dormandy J.A. (1987) Leukocytes and the risk of ischemic diseases. JAMA 257:2318-2324.

THE INTIMAL NEOVASCULARIZATION IN ATHEROSCLEROTIC PLAQUES OF HUMAN CORONARY ARTERIES

K.SUEISHI, M.KUMAMOTO, H.SAKUDA, Y.NAKASHIMA AND Y.ISHII
Department of Pathology, Faculty of Medicine, Kyushu University,
3-1-1 Maidashi, Higashi-ku, Fukuoka 812, Japan

INTRODUCTION

Although the intimal neovascularization, the newly formed vasa vasorum, in atherosclerotic plaques of human coronary arteries has been recognized for many years[1,2], there are still unresolved questions regarding the origin and pathogenesis of these newly formed blood vessels. It is well known that neovascularization is an ubiquitous and vital response in various physiological and pathological conditions such as embryonic development, inflammatory-repair process and growth of solid cancer[3]. The regulatory mechanism of neovascularization has been recently paid much attention and studied in some details from aspects on molecular biologies, but there have been disagreements and controversies concerning the pathobiological role and significance in vivo of many angiogenic factors.

The purposes of this paper are to manifest the morphological characteristics of newly formed blood vessels in atherosclerotic plaques of human coronary arteries, especially the origin of vascular ingrowth into the plaques and the association of intimal neovascularization with intimal hemorrhage and rupture, inflammatory cell infiltrate and accumulation of atheromatous gruel, and to clarify the effect of hypoxic state on smooth muscle cell (SMC) modulation of angiogenic process in vitro.

PATHOPHYSIOLOGICAL SIGNIFICANCE OF NEOVASCULARIZATION IN HUMAN CORONARY ATHEROSCLEROSIS

In cardiovascular lesions, neovascularization occurs ubiquitously as organization and recanalization of thrombi and aberrant vasa vasora in atherosclerotic intima, and in the reparative stage of angiitis. Several reports have found that newly formed microvessels in atherosclerotic intima of human coronary arteries are generally fragile and tortuous, suggesting that these vessels are related to the occurrence of intimal hemorrhage and rupture of atheroma, followed by formation of occlusive thrombi and vascular spasm as the occasionally fatal sequence. Acute myocardial infarction has been clinically shown to be largely caused by coronary thrombosis[4]. Examining the occurrence of coronary thrombosis and histopathological alterations of thrombotic sites of coronary arteries obtained at serial 84 autopsies of acute myocardial infarction, by postmortem angiographies with barium

---

Key words: Neovascularization, Atherosclerosis, Coronary artery, Transforming growth factor β, Smooth muscle cell

sulfate infusion method and light microscopic observation of successive sections, about 70% of these patients were accompanied with thrombus formation in epicardial coronary arteries perfusing the infarcted myocardium. The comparison between the types of acute myocardial infarction and prevalence of coronary thrombosis revealed the higher incidence of thrombi in massive type of acute myocardial infarction, especially in transmural type (90.2%), than in others (Table 1). The characteristics of histological alterations at the thrombotic sites of coronary arteries are shown in Table 2. Thrombi were always observed in the markedly obliterated coronary arteries more than 75% of cross-sectional area, due to atherosclerotic plaque. The main lesions were rupture of atheroma and intimal hemorrhage. These lesions were observed more than 70% of thrombotic sites and these thrombi were frequently occlusive (Table 2).

Table 1 Correlation between Types of Acute Myocardial Infarction and Coronary Thrombosis

Type	Case	Thrombosis	
Massive	54	46	(85.2%)
Transmural	41	37	(90.2%)
Not transmural	13	9	(69.2%)
Scattered	9	6	(66.7%)
Subendocardial	12	2	(16.7%)
Subtotal	75	54	(72.0%)
Undetermined	9	-3	(33.3%)
Total	84	57	(67.9%)

Table 2 Histological Changes at The Site of Coronary Thrombosis

Histological changes	Thrombus	Occlusive
Rupture of atheroma	31(50.0%)	23/32(74.2%)
Hemorrhage in the intima	14(22.6%)	8/14(57.2%)
Marked sclerosis, only	17(27.4%)	5/17(28.4%)
Total	62(100%)	36/62(50.0%)

Analyzing the origin of angiogenic growth from either luminal or adventitial vascular endothelial cells about 25 patients (average age: 74 yrs.) died of non-cardiac causes and 5 patients (average age: 64 yrs.) of cardiac causes, using postmortem angiographies of methyl salicylate-cleared hearts with silicone polymer, we segmentally documented delicate vascular networks or plexuses in close vicinity to the proper lumen of coronary arteries associated with atherosclerotic narrowing (Fig. 1). The light microscopic examination of successive sections, obtained from about 70 paraffin-embedded blocks with approximately 2 mm-thickness,

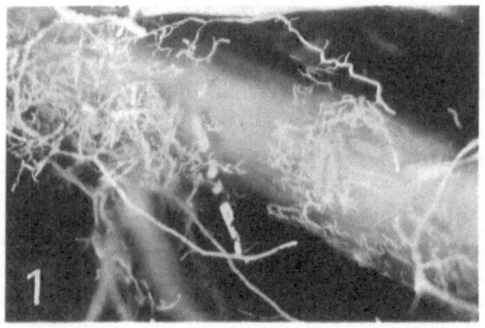

Fig. 1. Neovascularization of coronary artery. The newly formed vascular networks are segmentally documented with silicone polymer infusion, and locate in the close vicinity to the proper lumen of coronary artery with atherosclerotic narrowing.

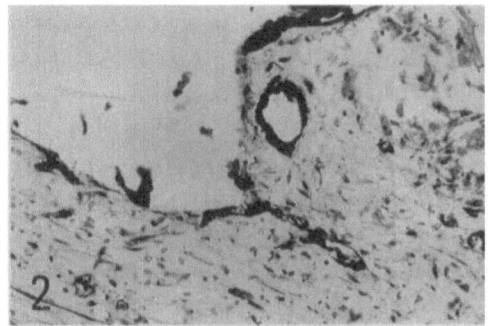

131

revealed that newly formed blood vessels in atherosclerotic lesions were mainly derived from adventitial and partly from luminal endothelial growth (Fig. 2).

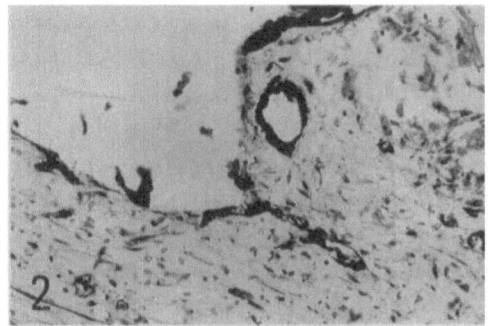

Fig. 2. Light microscopic finding of newly formed vessels in the atherosclerotic intima. A blood vessel directly connects with the proper lumen of coronary lumen, suggesting that this vessel in the intima is derived from luminal endothelial growth. H.E. x 580.

The extent of intimal neovascularization was relatively correlated with the severity of atherosclerosis, especially with luminal narrowing, inflammatory cell infiltrate, formation of atheroma and granulation-like tissue, and intimal hemorrhage, but not with calcification and hyalinization. The distribution of these blood vessels was not simple and varied in each patient, and this fact appeared to be fundamentally related to the variety of human atherosclerotic lesions, which might be progressed and/or regressed during a long period. However, the newly formed vessels derived from adventitial endothelial growth tended to localize in the shoulder portion of atherosclerotic plaque as well as in the deeper portion of atheroma, and, on the contrary, the blood vessels from luminal endothelial growth distributed mainly in the superficial intima as forming vascular plexuses, especially associated with formation of granulation-like tissue and chronic inflammatory infiltrate. Fibrin deposits and accumulation of hemosiderophages were present, but not frequently. These findings offer more evidences to support the hypothesis that there are three possible mechanisms of intimal neovascularization, namely, the metabolic explanation, thrombogenic and inflammatory theories. The metabolic explanation is that newly formed blood vessels result from the stimulation of adventitial endothelial cells to grow into plaques by local hypoxia[5,6]. The thrombogenic and inflammatory theories possess the many similarities that exist between organization of mural thrombi[7] and inflammatory cell infiltrate mainly composed of macrophages and T-lymphocytes[8], and atherosclerotic processes.

Intimal neovascularization is ubiquitously found in human atherosclerotic coronary arteries and probably a normal response to injury occurring during atherosclerotic processes. This angiogenic processes may involve various factors such as platelet-, macrophage-, fibroblast- and SMC-derived growth factors (PDGF, bFGF, ECDGF) and other modulating factors including TGFβ. Therefore, intimal neovascularization may be essential as an inflammatory-repair process in both progression and regression of coronary atherogenesis and the following sequelae, such as intimal hemorrhage, rupture of fibrous plaque and formation of occlusive thrombi, possibly leading to acute myocardial infarction. The latter sequelae may be induced by additional effects such as coronary spasm, hypertension or abrupt fluctuation of blood pressure on newly formed blood vessels in atherosclerotic intima.

ANGIOGENESIS IN VITRO: ENDOTHELIAL CELL - SMOOTH MUSCLE CELL INTERACTION

Atherosclerotic lesions are characterized by SMC proliferation, chronic inflammatory cell infiltrate and extra- and intracellular deposition of lipids. Intimal neovascularization, as described above, has been shown to correlate relatively to luminal stenosis in atherosclerotic leison in vivo, and has been also speculated to be regulated by the cell-to-cell interaction in vivo, namely, the interaction between endothelial cells and other cells such as SMCs and infiltrating inflammatory cells, especially macrophages and lymphocytes. Therefore, we examined the hypoxic influence to the modulating effect of SMCs on angiogenesis, using the quantitative assay method of in vitro angiogenesis model[9].

The bovine capillary endothelial cells (BCEs) were isolated from the bovine adrenal cortex[10] and eighth to tenth passage of culture was used in the following studies. The bovine SMCs were isolated from bovine aortic media by explant method[11] and used at sixth to 13th passage. Confluent SMCs (ca. 9.2×10^4 cells/cm^2) were cultured under the condition of 20, 5 or 1% O_2 balanced with N_2 in a humidified 0.5% CO_2 incubator at 37°C (Multigasincubator BL-3200, Astec Co., Fukuoka, Japan). The conditioned media were harvested after several hours (3-48 hrs.). BCEs (6×10^4 cells/well) were seeded on Type I collagen gels (0.2%) on the filter of the Millicells (12 mm in diameter, Millipore Co., Bedford, MA), which were inserted in 12-well plates (Corning Glass Works, Corning, NY). Conditioned media were applied to the outer well, and then BCEs in the Millicells were cultured in a conventional CO_2 incubator at the O_2 concentration of 20%. The extent of angiogenesis was assessed by measuring and calculating morphometrically the total length of tubular structures (mm/cm^2) organized by BCEs in Type I collagen gels[9].

The effect of SMC conditioned media on angiogenesis is shown in Table 3. SMC conditioned media enhanced the angiogenesis as compared with non-conditioned fresh culture medium as a negative control, and bFGF at the concentration of 10 ng/ml as a positive control. These enhancing effect of SMC conditioned media on angio-genesis was apparently dependent on the degree of hypoxic state. When anti-bFGF or TGFβ_1 polyclonal IgG, at the 10-timed concentration of 50% inhibition of respective effect of bFGF or TGFβ_1 at the concentration of 1ng/ml on endothelial growth, was added in this assay system for checking the effect of SMC conditioned media cultured at 1% O_2 concentration, anti-TGFβ_1 IgG significantly suppressed the angiogenic effect of SMC conditioned media, but anti-bFGF did not (Table 4).

Table 3 Tube formation of bovine capillary endothelial cells in the conditioned media of smooth muscle cells (SMC-CM).

Culture condition	Total length of tube (mm/mm^2)
Fresh medium	0.343 ± 0.170
SMC-CM	
20% O_2	0.739 ± 0.155*
5% O_2	1.673 ± 0.322**
1% O_2	2.543 ± 0.073***
bFGF(10ng/ml)	1.580 ± 0.220*

Mean values ($\pm$SD) of three experiments are shown.
* $p<0.05$, ** $p<0.01$, *** $p<0.001$ vs control

Table 4 Effects of anti-IgG against TGF-β1 or bFGF on tube formation of bovine capillary endothelial cells in conditioned of smooth muscle cells (SMC-CM).

Culture condition	Total length of tube (mm/mm^2)
Fresh medium	0.32 ± 0.08*
SMC-CM	
CM+non-immune IgG(30ug/ml)	2.49 ± 0.57
CM+anti-bFGF IgG(30ug/ml)	2.15 ± 0.54^{ns}
CM+anti-TGF-β IgG(30ug/ml)	0.67 ± 0.19*

Mean values ($\pm$SD) of three experiments are shown.
* $p<0.01$, ns (not significant) vs CM+non-immune IgG

In addition of in vitro assay of angiogenic reagents, we confirmed the real roles of possible angiogenic factors in vivo assay system. Fig. 3 shows the result of rabbit cornea assay method on 5th day after applying the stimulant. The polymer mesh containing serum-free SMC conditioned media at 1% O_2, which was concentrated and dialyzed with PBS, was embedded into the cornea. The newly formed blood vessels grow into the embedded portion of CM from the congested vascular networks at corneal limbus. No apparent neovascularization could be found in the negative control using polymer mesh with PBS. Fig. 4 shows the light microscopic finding of rabbit cornea containing newly formed blood vessels composed of veins and a small-sized muscular artery, and minimal inflammatory cell infiltrate mainly composed of macrophages. The extent of inflammatory cell infiltrate was not so strong and was not so different from that in the case of application of bFGF.

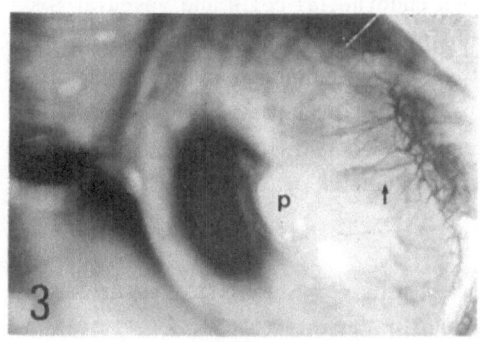

Fig. 3. Angiogenesis in the rabbit cornea by conditioned medium of smooth muscle cells clutured at a hypoxic state. Smooth muscle cells were cultured in serum-free DMEM at 1% O_2 for 24hr. and the conditioned medium was concentrated and dialyzed with PBS. The newly formed blood vessels (arrow) are developed to the intracorneal polymer mesh (P), containing smooth muscle cell-conditioned medium, from the corneal limbus.

Fig. 4. Light microscopic finding of the rabbit cornea shown in Fig. 3. There are newly formed veins and a small-sized muscular artery, associated with minimal inflammatory cell infiltrate mainly composed of macrophages. H.E. x 100.

Examining the effect of SMC conditioned media on thymidine incorporation by BCEs, SMCs and fibroblasts in the two-dimensional and conventional culture system without gel matrix, these thymidine incorporations were suppressed by SMC conditioned media and this suppression depended on hypoxic state. This inhibitory effect of SMC conditioned media on growth of these cells was also apparently abolished by simultaneous addition of anti-TGFβ_1 IgG, but not by anti-bFGF or PDGF IgG. These findings suggest that SMCs at hypoxic state release an active angiogenic factor(s) into conditioned media and active TGFβ_1 is closely related to the expression of the angiogenic effect by SMCs cultured under hypoxic state. Several studies[12,13] have demonstrated that TGFβ is released by various cells in a biologically latent form. This latent one is unable to bind to the TGFβ_1 receptor[14] and the activation mechanism of latent form to active one in vivo has been still unclear. Examining the effect of simultaneous addition of Glu-plasminogen to SMC conditioned media at 1% O_2 on angiogenesis, the angiogenic activity was enhanced by plasminogen in a dose-dependent manner. Furthermore, aprotinin suppressed this enhancing activity of plasminogen, suggesting that the angiogenic activity expressed by SMCs cultured under hypoxic state might be partly related to the activation activity of latent TGFβ to active one by unknown proteases produced by SMC.

Although I proposed the possible mechanisms concerning the intimal neovascularization in atherosclerotic lesions, these fingings offer more supportive evidences to the hypothesis that the interaction between endothelial cells and vascular SMCs seems to be very important in angiogenic processes even in atherosclerotic processes, and that active TGFβ participates in the enhancement and regulation of angiogenic processes in a paracrine system. The further studies, however, are necessary to prove the localization of active TGFβ in situ of atherosclerotic lesions and also to clarify the activation mechanisms of a latent TGFβ under hypoxic state.

This work was supported in part by Grants-in-Aids for Scientific Research from the Ministry of Education, Science and Culture of Japan (#63304045, 02219105).

REFERENCES

1. Winternitz M.C., Thomas R.M. and Le Compte P.M. (1938) The biology of athero-sclerosis. Springfield, Charles C Thomas.

2. Barger A.C., Beeuwkes R.III, Lainey L.L. and Silverman K.J. (1984) Hypothesis: Vasa vasorum and neovascularization of human coronary arteries. N. Engl. J. Med. 310: 175-177.

3. Folkman J. (1990) What is the evidence that tumors are angiogenesis dependent? J. Natl. Cancer Inst. 82: 4-6.

4. DeWood M.A., Spores J., Notske R., Mauser L.T., Burroughs R., Golden M.S. and Lang H.T. (1980) Prevalence of total coronary occlusion during the early hours of transmural myocardial infarction. N. Engl. J. Med. 303: 897-902.

5. Osborn G.R. (1963) The incubation period of coronary thrombosis. London, Butterworths.

6. Zemplenyi T., Crawford D.W. and Cole M.A. (1989) Adaptation to arterial wall hypoxia demonstrated in vivo with oxygen microcathodes. Atherosclerosis 76: 173-179.

7. Duguid J.B. and Robertson W.B. (1957) Mechanical factors in atherosclerosis. Lancet 1: 1205-1209.

8. van der Wal A.C., Das P.K., van der Berg D.B., van der Loos C.M. and Becker A.E. (1989) Atherosclerotic lesions in humans. In situ immunophenotypic analysis suggesting an immune mediated response. Lab. Invest. 61: 166-170.

9. Yasunaga C., Nakashima Y. and Sueishi K. (1989) A role of fibrinolytic activity in angiogenesis: Quantitative assay using in vitro method. Lab. Invest. 61: 698-704.

10. Folkman J., Haudenschild C.C. and Zetter B.R. (1979) Long-term culture of capillary endothelial cells. Proc. Natl. Acad. Sci. USA 76: 5217-5221.

11. Ross R. (1971) The smooth muscle cell. II. Growth of smooth muscle in culture and formation of elastic fibers. J. Cell Biol. 50: 170-186.

12. Laurence D.A., Pircher R., Kryceve-Martinerie C. and Jullien P. (1984) Normal embryo fibroblasts release transforming growth factors in a latent form. J. Cell. Physiol. 121: 184-188.

13. Assoian R.K., Fleurdelys B.E., Stevenson H.C., Miller P.J., Madtes D.K., Raines E.W., Ross R. and Sporn M.B. (1987) Expression and secretion of type beta transforming growth factor by activated human macrophages. Proc. Natl. Acad. Sci. USA 84: 6020-6024.

14. Wakefield L.M., Smith D.M., Masui C., Harris C. and Sporn M.B. (1987) Distribution and modulation of the cellular receptor for transforming growth factor-beta. J. Cell Biol. 105: 965-975.

OXYGEN FREE RADICALS INHIBIT Ca^{2+}-PUMP ACTIVITY BY OXIDIZING SULFHYDRYL GROUPS IN RAT HEART SARCOLEMMAL MEMBRANE.

MASANORI KANEKO, AKIRA KOBAYASHI, and NOBORU YAMAZAKI.

Third Department of Internal Medicine, Hamamatsu University School of Medicine, Hamamatsu, Japan.

INTRODUCTION

Although oxygen free radicals have been implicated as mediators of cellular injury in some pathophysiological conditions like myocardial ischemia-reperfusion [1], the mechanisms of myocardial cell damage caused by oxygen free radicals are still unclear. In this regard, it should be pointed out that oxygen free radicals are known to attack molecules of major biological significance and these include phospholipids, protein, and nucleic acids [2]. Since sulfhydryl groups (SH groups) are known to regulate the membrane-bound enzyme activities in the cell [3, 4, 5, 6, 7], it is possible that the oxidation of SH groups in the membrane may lead to the depression of enzyme activities because of oxygen free radicals [8].

Since intracellular Ca^{2+} overload has been considered to play a crucial role in ischemia-reperfusion injury [8, 9], it is possible that free radicals may damage myocardial membranes to promote the entry and/or inhibit the efflux of Ca^{2+} in the cell. The present study was undertaken to examine the effects of oxygen free radicals on Ca^{2+}-stimulated ATPase activity which is intimately involved in the extrusion of Ca^{2+} across the cell membrane, and SH groups in rat cardiac sarcolemmal membranes, in vitro.

MATERIALS AND METHODS

Isolation of sarcolemmal membranes;

Male Sprague-Dawley rats weighing 200-250g were killed by decapitation. Hearts were removed, and the ventricular tissue was processed for the isolation of sarcolemmal membranes by the method of Pitts [10]. Several marker enzymes study revealed that the purified sarcolemmal fraction employed in this study contained minimal contamination by other organelles [11].

Measurement of Mg^{2+}-ATPase and Ca^{2+}-stimulated ATPase activities;

For the estimation of Mg^{2+}-ATPase, sarcolemmal vesicles (20-40 μg protein) were preincubated at 37℃ for 5 min in 0.5 ml of medium containing 140mM KCl-10mM MOPS-Tris (pH7.4), 2mM $MgCl_2$ 5mM sodium azide, and 0.2mM ethylene glycol-bis (β-aminoethyl ether)-N, N,N',N'-tetraacetic acid (EGTA). The reaction was started by the addition of 4mM Tris ATP (pH7.4) and was terminated 5 min later with 0.5ml of cold 12% trichloroacetic acid; the liberated phosphate was measured by the method of Tausky and Shorr (11). Estimation of total (Ca^{2+}-stimulated + Mg^{2+}) ATPase was carried out in the medium containing 140mM KCl-10mM MOPS-Tris (pH7.4), 2mM $MgCl_2$, 5mM sodium azide, and 1×10^{-5} M free Ca^{2+} . The concentration of free Ca^{2+} in the medium was adjusted by using EGTA(11). The Ca -stimulated ATPase activity was the difference between the total ATPase and Mg^{2+} -ATPase activities. All these experimental conditions were the same as reported elsewhere(11).

ATP-dependent Ca^{2+} accumulation assay;

Sarcolemmal vesicles (100 μg protein) were preincubated in 0.5ml of medium containing 140mM KCl-10mM MOPS-Tris (pH7.4), 2mM $MgCl_2$, and desired amount of $^{45}CaCl_2$ -EGTA ($^{45}CaC l_2$39.56 mCi/ng; New England Nuclear, Boston, MA) to give 1x 10^{-5}M free Ca^{2+} (11). Ca^{2+} accumulation was initiated by adding 4mM Tris-ATP (pH7.4). After 5 min of incubation at 37℃, 250 μl aliquots were immediately filtered through Millipore filters (0.45 μm), washed with 2x 3-ml ice-cold KCl-MOPS and 1mM $LaCl_3$ (pH7.4), dried, and the radioactivity determined for calculating the total Ca^{2+} accumulation. Nonspecific Ca^{2+} binding was measured in the absence of ATP for each set of experiments. The ATP-dependent Ca^{2+} accumulation was calculated by subtracting nonspecific Ca^{2+} binding from total Ca^{2+} accumulation.

Determination of sulfhydryl group content;

Sulfhydryl content of the sarcolemmal membranes was determined with DTNB, according to the procedure described elsewhere (11). The assay system contained 20mM imidazole-HCl buffer (pH7.0), 1mM EDTA, and 0.2mM DTNB. For the determination of free sulfhydryl groups, 200 μg membrane protein was incubated for 3 min in the above medium, and the absorbance at 412 nm was measured. For total sulfhydryl groups determination, the membrane was incubated for 30 min in the assay medium in the presence of 0.1% sodium dodecyl sulfate (SDS). It should be mentioned that SDS dissociates the membrane and allows the measurement of both extra and intramembranal sulfhydryl groups. Calculation of the sulfhydryl groups was based on a molar extinction coefficient of 13,600 M/cm at 412 nm for the thiophenol reaction product. This was verified with cysteine as a standard.

Free radical-generating systems;

Superoxide anion radicals were generated by the xanthine oxidase (Calbiochem) reaction in which xanthine was used as a substrate[2]. Superoxide dismutase (SOD; Sigma), was used as a scavenger for superoxide anion radicals.

Statistical analysis;

Results are presented as mean $\pm$SE. For statistical evaluation, multiple analysis of variance was carried out, and Duncan's multiple-range test was used to determine differences between the means within the population. $P<0.05$ was taken to reflect a significant difference.

RESULTS
Effects of xanthine plus xanthine oxidase on ATP-dependent Ca^{2+} accumulation and ATPase activities;

The effects of superoxide radicals on ATP-dependent Ca^{2+} accumulation and ATPase activities were examined by incubating the sarcolemmal membranes in the presence of 2mM xanthine plus 0.03U/ml xanthine oxidase. Figure 1 shows that the incubation of membranes with either 2mM xanthine, 0.03U/ml xanthine oxidase, or 80 μg/ml SOD, for 30 min did not affect ATP-dependent Ca^{2+} accumulation and ATPase activities. However, a combination of xanthine plus xanthine oxidase depressed ATP-dependent Ca^{2+} accumulation, and Ca^{2+} -stimulated ATPase activities by 65% and 81% of the control values, respectively. SOD showed a protective effect on the depression in Ca^{2+} accumulation and ATPase activities due to incubation with xanthine plus xanthine oxidase. To study the time-course effects of xanthine plus xanthine oxidase on Ca^{2+} -stimulated ATPase activity, the enzyme activities were measured after 1, 5, 10, 30, and 60 min preincubation at 37℃. Ca^{2+} -stimulated ATPase activity was inhibited in a time-dependent manner. Sarcolemmal membranes were also incubated without xanthine plus xanthine oxidase for the same periods, and these enzyme activities did not show any significant change even after 60 min (data not shown). These results indicate that the interaction of superoxide radicals with sarcolemmal membrane vesicles was rapid, because significant inhibition of Ca^{2+} -stimulated ATPase activity was seen after 1 min of incubation.

Effects of oxygen free radicals on sulfhydryl groups;

To assess the effects of superoxide radicals on membrane sulfhydryl groups, sarcolemmal membranes were incubated with xanthine plus xanthine oxidase system for 10 min at 37℃. The data in Figure 2 indicate that superoxide radicals significantly depressed both free and total sulfhydryl groups. This depression of sulfhydryl groups by superoxide radicals was lessened by the addition of SOD. Xanthine, xanthine oxidase, or SOD did not show any significant effects on sulfhydryl groups.

To demonstrate that sulfhydryl groups play an important role in the manifestation of Ca^{2+} -stimulated ATPase activity of the heart sarcolemmal membranes, the effects of N-ethylmaleimide (NEM) on Ca^{2+} -pump activities were investigated. Ca^{2+} -stimulated ATPase activity was depressed by NEM in a concentration-dependent manner (data not shown). Minimum concentration of NEM whichexhibited a significant effect on Ca^{2+} -stimulated ATPase activity was 10 μM. NEM also showedinhibitory effects on Ca^{2+} -stimulated ATPase activity in atime-dependent manner; a significant depression of Ca^{2+} -stimulated ATPase activity by 25 μM NEM was seen after 5 minincubation. The inhibitory effects of NEM on ATP-dependent Ca^{2+} accumulation and Ca^{2+} -stimulated ATPase were blocked by the addition of 1mM dithiothreitol (DTT) or 10mM cysteine.

Effects of sulfhydryl group-reducing agents on depression in ATP-dependent Ca^{2+} accumulation and Ca^{2+} -stimulated ATPase activities because of oxygen free radicals;

The incubation of sarcolemmal membranes with superoxide radicals generating system (2mM xanthine plus 0.03U/ml xanthine oxidase) for 10 min at 37℃ showed a significant depression of ATP-dependent Ca^{2+} accumulation and Ca^{2+} -stimulated ATPase activities (Figure 3). It can be

Figure 1: Effects of xanthine + xanthine oxidase on ATP-dependent Ca²⁺ accumulation and ATPase activities.

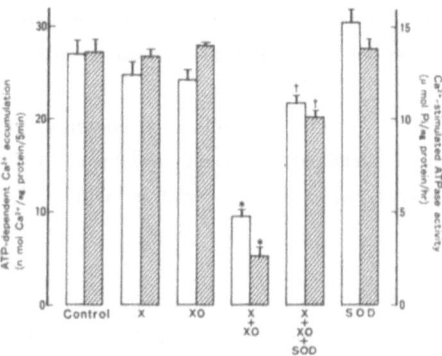

Prior to assay for ATP-dependent Ca²⁺ accumulation (☐) and ATPase activity (▨), sarcolemmal membranes (200 μg protein／ml for Ca²⁺ accumulation and 300-400 μg protein／ml for ATPase activity) were incubated for 30 min at 37℃ in 140 mM KCl, 10 mM MOPS, pH 7.4 plus the additions shown. The final concentrations of xanthine (X), xanthine oxidase (XO), and superoxide dismutase (SOD) were 2 mM, 23 μg／ml (0.03 U／ml), and 80 μg／ml, respectively. Each value is a mean± S.E. of 6 different preparations.
＊significantly different from control values (P＜0.05).
┼significantly different from X+XO values (P＜0.05).

Figure 2: Effects of xanthine + xanthine oxidase on heart sarcolemmal sulfhydryl groups.

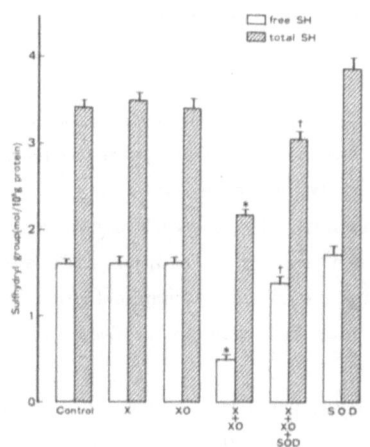

Prior to assay for sulfhydryl groups, sarcolemmal membranes (1 mg protein／ml) were incubated for 10 min at 37℃ in a medium containing 10 mM Tris - HCl, pH 7.4 plus the additions shown. The final concentrations of xanthine (X), xanthine oxidase (XO), and superoxide dismutase (SOD) were 2 mM, 23 μg／ml (0.03 U／ml), and 80 μg／ml, respectively. Each value is a mean ±S.E. of 6 different preparations.
＊significantly different from control values (P＜0.05).
┼significantly different from X+XO values (P＜0.05).

Figure 3: Effects of DTT or cysteine on the depression in ATP-depndent Ca accumulation and ATPase activities due to oxygen free radicals.

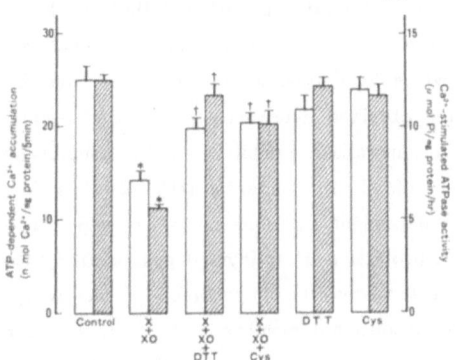

Prior to assay for ATP-dependent Ca²⁺ accumulation (☐) and ATPase activity (▨), sarcolemmal membranes (200 μg protein／ml for Ca²⁺ accumulation and 300-400 μg protein／ml for ATPase activity) were incubates for 10 min at 37℃ in 140 mM KCl, 10 mM MOPS, pH 7.4 plus the additions shown. The final concentrations of xanthine (X), xanthine oxidase (XO), dithiothreitol (DTT), cysteine (Cys) were 2 mM, 23 μg／ml (0.03 U／ml), 1 mM, 10 mM, respectively. Each value is a mean± S.E. of 6 different preparations.
＊significantly different from control values (P＜0.05).
┼significantly different from X+XO values (P＜0.05).

seen that the inhibition of Ca^{2+} -pump activities by superoxide radicals was significantly blocked by the addition of 1mM DTT or 10mM cysteine. DTT shows the protective effects on changes in Ca^{2+} -stimulated ATPase activity because of xanthine plus xanthine oxidase in a dose-dependent manner. It may be noted that 0.25mM or higher concentration of DTT was required to exert a significant effect on this system.

To find a relationship between Ca^{2+} -stimulated ATPase activity and sulfhydryl contents, the data from experiments concerning the effects of oxygen free radicals on heart sarcolemma were plotted. Both free and total sulfhydryl groups showed a significant correlation with Ca^{2+}-stimulated ATPase activity.

DISCUSSION

Incubation of cardiac sarcolemmal membranes with xanthine plus xanthine oxidase resulted in a dramatic inhibition of both ATP-dependent Ca^{2+} accumulation and Ca^{2+} -stimulated ATPase activity. The reaction of superoxide radicals, generated by the xanthine plus xanthine oxidase system, with the membrane vesicles was rapid as 15% inhibition of the enzyme activity was seen after 1min of incubation.

Sulfhydryl groups are considred essential for proper functioning of the membrane-bound enzymes [3, 4, 5, 6, 7]. In this regard, the activities of kidney Na^+ -K^+ -ATPase [4] and skeketal muscle SR [6, 7], as well as liver plasma membrane Ca^{2+} -stimulated ATPase [12], were reduced by sulfhydryl reagents NEM and diamide. In the present study, we have shown that Ca^{2+}-stimulated ATPase activity and ATP-dependent Ca^{2+} accumulation of the heart sarcolemmal membrane were inhibited by NEM in a dose- and time-dependent manner. In addition, DTT and cysteine were found to exert protective effects on the depression in Ca^{2+}- pump activities by NEM. These results support the view that sulfhydryl groups may be intimately involved in the regulation of Ca^{2+} -pump activity in the sarcolemmal membrane.

Treatment of the heart sarcolemmal membranes with oxygen free radicals showed a significant depression in the superficial sulfhydryl groups (free sulfhydryl groups) and intramembranous sulfhydryl groups (total minus free sulfhydryl groups). These results suggest that oxygen free radicals can react and modulate not only the superficial but also the intramembranous portion of the membrane-bound enzyme protein. It is pointed out that the heart sarcolemmal membrane contains several types of enzyme proteins, such as Na^+ -K^+ -ATPase, Na^+ -Ca^{2+} exchanger, and Ca^{2+} -stimulated ATPase, which have been shown to be affected by oxygen free radicals [13, 14]. Thus it appears that the observed depression in the Ca^{2+} -pump activities by oxygen free radicals may be due to a generalized inhibition of the membrane sulfhydryl groups. It was interesting to observe that sulfhydryl reductants, DTT and cysteine, showed protective effects on the depression in ATP-dependent Ca^{2+} accumulation and Ca^{2+} -stimulated ATPase activities by oxygen free radicals. A protection of sulfhydryl groups from the oxidative stress because of free radicals further substantiates the importance of sulfhydryl groups in maintaining the Ca^{2+} -pump activity in the sarcolemmal membrane. In fact, a linear relationship between changes in the sarcolemmal Ca^{2+} -pump activity and sulfhydryl groups because of free radicals was observed in this study.

The data presented in this study indicate that when cardiac sarcolemmal membrane was exposed to oxygen free radicals, Ca^2 -pump activity was reduced. Such a defect can be seen to decrease Ca^{2+} extrusion from the myocardium.

REFERENCES

1. Ambrosio G, Becker LC, Hutchins GM, Weisman HF, Weisfeldt ML (1986) Circulation 74:1424-1433

2. Hammond B, Hess ML (1985) J Am Coll Cardiol 6:215-220

3. Scherer NM, Deamer DW (1986) Arch Biochem Biophys 246:589-610

4. Schoot BM, Schoots AFM, Depont JJHHM, Schuurmans-stekhoven FMAH, Bonting SL (1977) Biochem Biophys Acta 483:181-192

5. Schuurman-stekhoven FMAH, Bonting SL (1981) Physiol Rev 61:1-76

6. Yamada S, Ikemoto N (1978) J Biol C hem 253:6801-6807

7. Yoshida H, Tonomura Y (1976) J Biochem 79:649-654

8. Freeman BA, Crapo JD (1982) Lab Invest 47:412-426

9. Halliwell B, Gutteridge JM (1982) Advances in studies on heart metabolism. Bologna, CLUEB, pp 403-411.

10.Pitts BJR (1979) J Biol Chem 254:6232-6235

11.Kaneko M, Elimban V, Dhalla NS (1989) Am J Physiol 257 (Heart Circ Physiol 26):H804-H811

12.Belomo GF, Mirabelli P, Rickelmi P, Orrenius S (1983) FEBS Lett 163:136-139

13.Kramer JH, Mak IT, Weglicki WB (1984) Circ Res 55:120-124

14.Reeves JP, Bailey CA, Hale CC (1986) J Biol Chem 261: 4948-4955

Clinical Studies

M.H. PIETRASZEK, Y. TAKADA AND A. TAKADA
Department of Physiology, Hamamatsu University School of Medicine, Shizuoka-ken, Hamamatsu-shi, 3600 Handa-cho, 431-31 Japan.

INTRODUCTION

Serotonin (5-hydroxytryptamine; 5HT) is found in the central nervous system, which contains serotonergic neurons, enterochromaffin cells in the intestine and platelets.[1-3] Receptors for serotonin have been found in various tissues including the central nervous system, gastrointestinal tract , blood vessels, blood platelets and autonomic nerve endings.[4] 5-hydroxytryptamine is believed to play a role in physiology and pathophysiology of mammals. One of its most likely physiological roles is to aid in haemostasis by promoting platelet aggregation and by causing local vasoconstriction.[2,3] It also has a role in some forms of vascular disease and may contribute to vasospasm of cerebral or coronary arteries, especially with endothelial disfunction or damage.[5] Some evidence has implicated serotonin in the pathogenesis of migraine, peripheral vascular disease, coronary and cerebral vasospasm or essential hypertension. On the other hand, clinical studies have provided evidence that serotonin in the brain is implicated in the state of depression. [6,7] Many authors consider that blood platelets share a number of properties with serotonergic neurons because of their specific biochemical mechanisms for uptake and storage of amine. [8,9] Thus, the study of blood serotonin would provide an useful model for certain aspects of neuronal physiology.

The present study investigates blood serotonergic mechanisms in diseases with blood vessels disfunction like Thromboangiitis obliterans (Buerger's disease), Raynaud's phenomenon, hypertension and diabetes mellitus as well as in mental disorders as depression and neurosis.

MATERIALS AND METHODS

Patients Twenty patients with Buerger's disease, ten with Raynaud's phenomenon, ten with borderline hypertension, 20 with diabetes mellitus , 20 with depression and 10 with neurosis participated in the present investigation. Table 1 shows their characteristic. All of patients were drug-free at least one week before experiment. The control group matched for age and sex with the patients was made up of 50 apparently healthy volunteers.
Blood collection Blood samples were taken in the morning from antecubital vein in 10 ml tube containing 1.0 ml of 3.8% sodium citrate. The blood was divided into two portions, one of which was added with ascorbic acid and was used for the determination of whole blood serotonin content. The other was immediately centrifuged at 200 g for 10 min to prepare platelet rich plasma

(PRP). Platelet poor plasma (PPP) was obtained by additional centrifugation ofthe blood at 1200 g for 20 min. The part of PPP was added with ascorbic acid and was used for estimation of free serotonin concentration. For 5-HT uptake and release P,RP was diluted with autologous PPP to give a stock suspention of 10^9 platelets/ ml.

Measurements Whole blood and plasma serotonin concentrations were analyzed in duplicate using the HPLC with fluorescence detector according to Anderson et al.[10]

The uptake and release reaction of labeled [^{3}H] serotonin was measured according to Lingjaerde.[11]

Statistical analysis Group differences were tested for significance using a Wilcoxon rank sum test. A value of $p<0.05$ was taken as a level of significance. Results are expressed as a mean ± standard error.

RESULTS

Table 1. The characteristic of studied population.

	number	sex F / M	age mean	range
Control group	40	12 / 28	35	22-58
Buerger's disease	20	0 / 20	42	32-67
Raynaud's phenomenon	10	0 / 10	37	33-56
Diabetes mellitus	20	6 / 14	47	39-67
Depression	20	10 /10	46	25-65
Neurosis	10	2 / 8	39	35-53

Whole blood concentration of serotonin
Fig. 1 summarizes the obtained results of whole blood serotonin concentration in studied groups of patients. Patients with Buerger's disease, Raynaud's phenomenon and depression had the lowest blood serotonin concentration when compared to remaining patients and healthy population. Diabetic and neurotic subjects showed also lower blood level of amine but in those cases decrease was not so pronounced as in vascular diseases or depression. Only the group of hypertensive patients showed no significant changes in blood serotonin when compared to control group. Whole blood concentration of 5HT in two vascular diseases - Buerger's disease and Raynaud's phenomenon did not differ significantly.

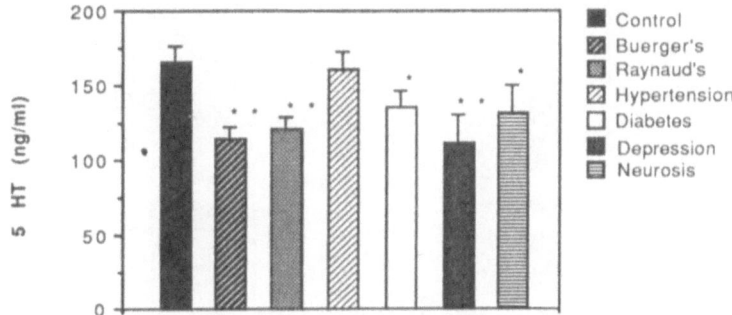

Fig.1. Whole blood serotonin concentration in studied groups of patients. **-p<0.01, *- p<0.05 in respect to control value. Results are expressed as a mean ± standard error.

Plasma level of serotonin

Plasma concentration of serotonin is depicted in Fig. 2. The highest concentration of free serotonin in plasma was observed in Buerger's disease and Raynaud's phenomenon. Patients suffering from diabetes mellitus had three fold higher concentration of amine in the plasma when compared to the control group. Plasma level of 5HT in hypertension was significantly elevated but it was less pronounced than those of other patients.

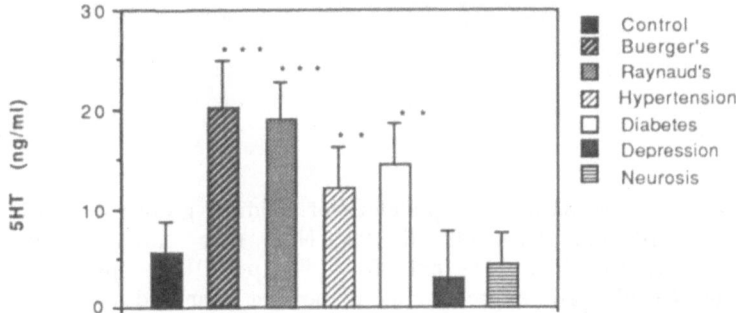

Fig.2. Plasma serotonin concentration in studied groups of patients. ***- p<0.001, ** - p<0.01, *- p<0.05 in respect to control value. Results are expressed as a mean ± standard error.

On the contrary plasma levels of serotonin in depression and neurosis were slightly (non-significantly) lower than that of controls. To explain above mentioned findings we investigated uptake of serotonin by blood platelets.

Uptake of labeled serotonin by platelets

As could be expected the lowest value of 5HT uptake was observed in group of patients suffering from depression (Fig.3). However it was not correlated with blood serotonin concentration (r=0.25. NS. data not shown).

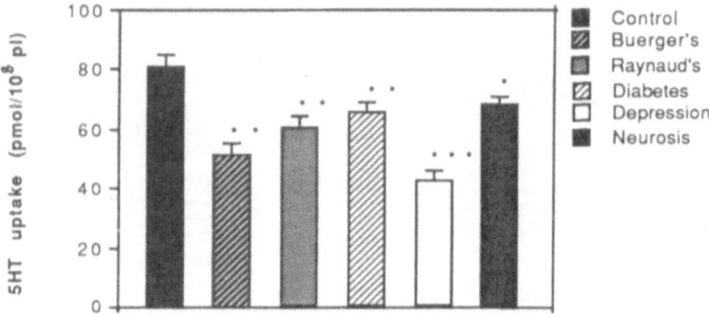

Fig.3. Uptake of serotonin by blood platelets in studied groups of patients. Note that in hypertension this experiment has not been done. ***- p<0.001, ** - p<0.01, *- p<0.05 in respect to control value. Results are expressed as a mean ± standard error.

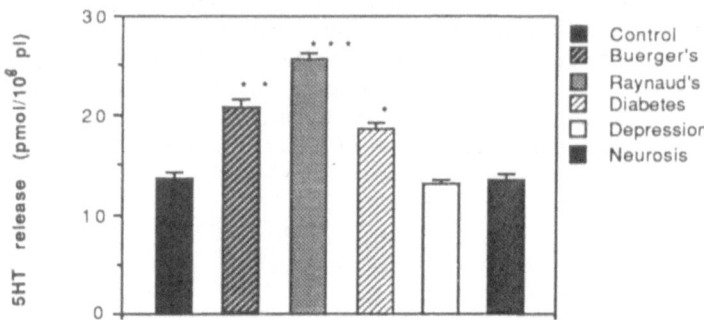

Fig.4. Release of serotonin from platelets of studied groups of patients. Platelet suspention was incubated for 60 minutes. Note that this experiment has not been done in hypertension. ***- p<0.001, ** - p<0.01, *- p<0.05 in respect to control value. Results are expressed as a mean ± standard error.

In Buerger's disease and Raynaud's phenomenon active transport of serotonin into the platelets was reduced by 25% and 20% respectively. These values were positively correlated with whole blood serotonin content (r=0.46, p<0.05 in Buerger's disease and r=0.51, p<0.05 in Raynaud's phenomenon). Patients suffering from diabetes and neurosis had decreased uptake by 12% and 15% respectively when compared with control group.

Release of labeled serotonin from platelets

Serotonin-loaded platelets were resuspended in the autologous PPP and incubated for 60 minutes. Radioactivity of released serotonin was counted in the medium. As seen in Fig.4 platelets from vascular diseases and diabetes released more serotonin than those of healthy subjects. In mental disorders release reaction was comparable to that of control group.

DISCUSSION

Our present results show clearly the differences in the blood serotonergic system in various diseases. Decrease in whole blood serotonin going together with an increase of its free fraction in the plasma were common in peripheral vascular diseases. Similar disarrangement was observed in diabetes, especially in patients with vascular complications. Hypertension was characterized by higher level of plasma serotonin. Finally, depression and neurosis were accompanied by decreased level of serotonin in whole blood, whereas amine concentration in the plasma remained unchanged in neurosis or even lower than that of control in depression. Decrease in blood serotonin concentration reflects lower than normal platelet serotonin content since very little of amine is detectable outside of the platelets.[2,3] Considering that platelets do not synthesize serotonin, the intraplatelet content of amine is likely to be dependent upon the net balance of its uptake and release. Our present data provide evidence that whole blood serotonin concentration depends on its transport activity into blood platelets. It was supported by very high correlation between blood serotonin concentration and uptake of amine by platelets. if some biochemical similarities between platelets and synaptosomes are taken into consideration the low platelet serotonin level in patients could simultaneously be accompanied by a decreased concentration of 5HT in the brain. It is probably valid in the case of depression only, since uptake of serotonin in depressed patients was the lowest. This is in agreement with the theory of endogenous depression as a consequence of serotonin deficiency in certain brain areas.[6] This theory, however is based on the lower 5HIAA concentration in cerebrospinal fluid of depressive patients and lower 5HT level in the brains of such patients after death.[7] The low level of blood serotonin in peripheral vascular diseases and diabetes does not reflect the concentration of amine in the brain, because such decrease is mainly due to enhanced release reaction. Increased level of free serotonin in the plasma speaks in favour of that statement. On the other hand enhanced release of serotonin with concomitant impaired uptake of amine may lead to the elevation in plasma free serotonin. Increase in plasma serotonin level shown in the groups of patients with Buerger's disease, Raynaud's phenomenon, hypertension and diabetes could result in a tendency to vasospasm, which is relevant to both structural vascular disease and secondary vasospastic disease. Serotonin present in excess in the plasma could lead to activation of platelets via $5HT_2$ receptors at the platelet membrane and the consequent further release of amine from platelets.[2] Serotonin itself is a very weak activator of human platelets, but in some pathological condition aggregation of platelets with degranulation have been shown. 5HT can also induce increase in intracellular messengers causing sensitization of platelets to aggregating agents.[12] Moreover, serotonin when released from activated platelets may induce the activation of peripheral vascular myofibroblasts. In such cases fibrosis and cell migration may produce obliterating fibrotic lesions in the vessel wall.

In conclusion we have shown that alterations of the platelet serotonergic system as impairment of 5HT transport and storage could be found in various pathological states including mental, neurological and cardiovascular disorders.

REFERENCES

1.   Da Prada M., Richards J.G. and Kettler R. (1981) Platelets in Biology and Pathology. Amsterdam: Elsevier: 107-45.

2.   Cohen R.A. and Vanhoutte P.M. (1985) Serotonin and the Cardiovascular System. New York: Raven Press : 105-12.

3.   Van Zwieten P.A. (1987) Pathophysiological relevance of serotonin. J. Cardiovasc. Pharmacol. 10 (Suppl.3):S19-S25.

4.   Peroutka S.J. (1988) 5-hydroxytryptamine receptor subtypes. Ann. Rev. Neurosci. 11: 45-60.

5.   Vanhoutte P.M. (1985) Serotonin and the cardiovascular system. New York: Raven Press :123-33.

6.   Cowen P.J. (1988) Recent views on the role of 5-hydroxytryptamine in depression. Curr. Opin. Psychiatry 3: 56-59.

7.   Coppen A.J. and Doogan D.P. (1988) Serotonin and its place in the pathogenesis of depression. J. Clin. Psychiatry 8 (suppl) : 4-11.

8.   Pletscher A. (1988) Platelets as models : use and limitations. Experientia 44: 152-155.

9.   Stahl S.M. (1978) The human platelet- A diagnostic and research tool for the study of biogenic amines in psychiatric and neurologic disorders. Arch. Gen. Psychiatry 34: 509-516.

10. Anderson G.M., Young J.G., Ohen D.J., Schlicht K.R. and Patel N. (1981) Liquid-chromatografic determination of serotonin and tryptophan in whole blood and plasma. Clin. Chem. 5: 775-6.

11. Lingjaerde O. (1979) Inhibitory effect of ethanol on 5-hydroxy-tryptamine (serotonin) uptake in human blood platelets in vitro. Acta. Pharmacol.Toxicol. 45: 394-8.

12. Pletscher A. (1987) The 5-hydroxytryptamine system of blood platelets : physiology and pathophysiology. Int. J. Cardiol. 14: 177-188.

INFLUENCE OF MENTAL STRESS ON FIBRINOLYSIS , PLATELET AGGREGATION AND AMINES IN PLASMA

Y. TAKADA, M.H. PIETRASZEK, T. URANO AND A. TAKADA
Department of Physiology, Hamamatsu University, School of Medicine, Hamamatsu-shi, Shizuoka-ken, Japan 431-31

INTRODUCTION

Many investigators are interested in the relationship between mental stress and platelet function, coagulation and fibrinolysis [1, 2], since several lines of evidence suggest that atherosclerotic cardiovascular events may be related to mental stress. Platelet aggregability in response to ADP increased in association with cardiac catheterisation and other diagnostic procedures [3], but on the contrary diminshed platelet aggregation in response to stress was observed by Haft and Arkel [4] and Larsson et al [5]. Levine et al found that mental stress provoked a rise in platelet release of platelet factor 4 and beta thromboglobulin [6]. Mental stress has been known to be capable of shortening the whole blood clotting time [7, 8]. Ogston et al found that recalcified plasma clotting time was shorter and plasma fibrinolytic activity was higher in patients having anxiety at the time of venepuncture [9]. The initial observation that the fear of impending surgery increased fibrinolytic activity was made by Macfarlane and Biggs [10] Recently Jern et al [11] reported that the effects of mental stress on plasma coagulation and fibrinolysis. They showed that von Willebrand factor antigen, factor VIII coagulant activity and factor VII caoagulant activity increased significantly in response to mental stress, and furthermore mental stress caused an activation of the fibrinolytic system with an elevation of t-PA activity and t-PA antigen.

In the present study, we employed the popular laboratory test of bleeding time (Duke method) as the initiator of mental stress, since the procedure itself is fearful for the examinee due to the fact that the examinee can not see the procedure of incision to ear directly. We analyzed the effects of mental stress on the fibrinolytic system, platelet aggregation and plasma levels of catecholamines and serotonin.

MATERIALS AND METHODS

Initiation of mental stress The laboratory test of bleeding time (Duke method [11]) was used to initiate anxiety. Volunteers of 41 male students in Hamamatsu University School of Medicine, who took the laboratory course of hematology, were subjected to the study. Students, who have never experienced the test of the bleeding time, were exposed to the test performed by other students, who have never experienced this test as an examiner. Control blood samples were obtained from volunteers in advance in a resting condition at 8:30 am. Students were then informed of the bleeding time experiments and were intentionally told that small incision might not have resulted in enough bleeding, so that fairly large incision would be needed for successful test of the bleeding time. Blood was taken before and after the bleeding time tests.

Volunteers of 77 male students were subjected to measure platelet aggregation, and plasma levels of catecholamines and serotonin. In this case blood was taken before the bleeding time test, 30 min and 7 hours after the bleeding time test.

Euglobulin clot lysis time (ELT)   ELT was measured using an automatic microtiter plate reader [13].   Briefly the turbidity of the wells, which decreased as the consequence of euglobulin clot lysis, was measured as a function of the absorbance at 340 nm every 30 min.   The time of the midpoint between maximum absorbance and minimum absorbance was adopted as ELT.

Plasma levels of fibrinolytic components   Tissue plasminogen activator (t-PA), urokinase(UK), total plasminogen activator inhibitor-1(PAI-1) and t-PA-PAI-1 complex were measured by enzyme immunoassay [14-16].   The concentration of free PAI-1 in the plasma was calculated by the subtraction of the concentration of tPA-PAI-1 complex from that of total PAI-1.

Plasma levels of amines   The plasma concentration of serotonin was measured by high performance liquid chromatography (HPLC) according to the method of Anderson et al [17].   Plasma catecholamine levels were determined by HPLC after liquid-liquid extraction using the method based on the procedure of Van der Horn et al [18].

Statistical analysis   Data are presented as mean ± SD.   Student paired t-test was used to evaluate the significance of intergroup differences observed.

RESULTS

Changes in fibrinolytic parameters after mental stress

Table 1 Changes in fibrinolytic parameters after mental stress

		control	before test	after test
ELT	(hrs)	6.70 ± 0.15	4.80 ± 0.40	4.22 ± 0.37 *
t-PA	(ng/ml)	4.67 ± 0.37	4.84 ± 0.43	5.00 ± 0.37
UK	(ng/ml)	not done	1.32 ± 0.28	1.09 ± 0.08
PAI-1(c)	(ng/ml)	3.82 ± 0.28	1.94 ± 0.25	2.04 ± 0.26
PAI-1(t)	(ng/ml)	10.43 ± 1.05	6.55 ± 0.88	6.10 ± 0.86
PAI-1(f)	(ng/ml)	6.61 ± 0.86	4.61 ± 0.67	4.06 ± 0.65 **

PAI-1(c): t-PA-PAI-1 complex, PAI-1(t): total PAI-1, PAI-1(f): free PAI-1
*$p<0.005$, **$p<0.05$ compared with before test

Table 1 shows the effects of mental stress on the fibrinolytic system.   ELT shortened significantly after the bleeding time test ($p<0.005$).   It is interesting that the significant shortening of ELT ($p<0.001$) was observed by merely giving verbal intimidation to volunteers.   Plasma t-PA levels after the bleeding time test were slightly higher than before the bleeding time test, and controls. Plasma UK level did not change significantly.   Total PAI-1 levels just before experiments were significantly shorter than those of controls ($p<0.001$).   Total

PAI-1 levels slightly decreased after the bleeding time test. t-PA-PAI-1 complex decreased significantly from control levels to the level of prior to the bleeding time test and the post bleeding time test. ($p < 0.001$). Plasma levels of free PAI-1 decreased significantly from control levels to levels of before bleeding time test ($p < 0.001$) and further to the levels of the post bleeding time test. From these results the enhanced activity of the fibrinolytic system is mostly due to the decrease in PAI-1, together with increase in t-PA.

Changes in plasma levels of serotonin, epinephrine and norepinephrine
Plasma levels of serotonin, epinephrine and norepinephrine increased 30 min after the test, however, 7 hours later it restored to the pre-test value. (Table 2)

Table 2   Plasma levels of amines after mental stress

		before test	30 min after test	7 hrs after test
serotonin	(ng/ml)	4.7 ± 2.8	6.4 ± 2.9*	4.8 ± 3.2
epinephrine	(pg/ml)	76 ± 9	132 ± 25*	85 ± 12
norepinephrine	(pg/ml)	318 ± 21	719 ± 34**	369 ± 23

*$p < 0.05$, **$p < 0.01$ compared with before test

Effects of mental stress on platelet aggregation

Table 3   Platelet aggregation after mental stress

agonist	before test	30 min after test	7 hrs after test
serotonin (1.0 μM)	(%) 38 ± 6	(%) 51 ± 5*	(%) 39 ± 7
ADP (4.0 μM)	74 ± 6	93 ± 6*	79 ± 8
collagen (2.0 μg/ml)	80 ± 5	84 ± 8	79 ± 7

Aggregation is expressed as the increase in light transmission.
*$p < 0.05$ compared with before test

Table 3 shows the results of the platelet aggregation induced by ADP, serotonin and collagen. Aggregation is expressed as increase in light transmission four minutes after the addition of agonists. 30 minutes after the bleeding time test platelets showed significantly greater responses to serotonin and ADP ($p < 0.05$), and slightly higher response to collagen. 7 hours after the test platelet responses returned to the pre-stress conditions.

Relationship between the bleeding time and the plasma levels of serotonin

In order to elucidate the possible association between the bleeding time and plasma amine concentrations we correlated two variables.   Serotonin in plasma showed a high inverse correlation with the bleeding time (r = -0.058, p<0.01).   Epinephrine in the plasma had weak, not statistically significant negative correlation with the bleeding time (r=-0.322, p<0.08).   Plasma norepinephrine level showed no correlation with the bleeding time.

DISCUSSION

The coagulation and fibrinolytic system is reported to be activated by mental stress [1, 2, 7-11].   It is also reported that platelet function is influenced by mental stress [1-6].   Since the laboratory test of bleeding time (Duke method) sometimes induces anxiety to patients because patients can not see the procedure of incision on the ear directly, we employed this procedure as the initiator of mental stress and reassessed its effect on fibrinolytic system and platelet function.   We intentionally stressed the fact to students that examiners were not qualified doctors but students who have never tried this test.   This information seemed to be enough stress to students from their attitudes and verbal expressions after having heard these comments.

We selected only male students as volunteers, since the circadian changes of the parameters in the fibrinolytic system in the morning are less in males than in females [19].

ELT, which is supposed to express the general activity of the fibrinolytic system, shortened significantly by mental stress.   This agrees with the previous data, which showed the increased levels of fibrinolytic activities after exposure to mental stress [1,2, 10, 11].   t-PA levels in the present study showed slight increase, although it was not statistically significant.   PAI-1 levels in the plasma, however, decreased after exposure to mental stress. Increase in t-PA after mental stress is explained by the stimulation by catecholamines, due to the fact that systemic pressure and heart rates are high during mental stress, and that the injection of adrenaline or physical exercises can initiate the similar fluctuation of parameters in the fibrinolytic system to that induced by mental stress [11].   In fact catecholamines and serotonin increased after mental stress (Table 2).   The increased level of catecholamines during mental stress may result in increase in t-PA release and in decrease in PAI-1 release from endothelial cells.   The mechanism of decrease of PAI-1 levels, however, is not clear yet.   Recently it is reported that [20] β-adrenergic agents stimulated the expression of t-PA gene and release of t-PA from cultured granulosa cells with concomitant increase in cyclic AMP contents, and that cyclic AMP potentiated phorbol ester stimulation of t-PA release, although it inhibited the secretion of PAI-1 from human endothelial cells [21].   This potentiation of t-PA release coupled with a reduction in PAI-1 secretion may have resulted in decrease in free PAI-1 levels and shortened ELT after mental stress.

Most interesting data may be the significant shortening of ELT and decreased plasma levels of PAI-1 (free and total) by merely giving verbal intimidation to volunteers.   These results suggest that volunteers were psychologically more influenced by the explanation of the test than the test itself.

As to platelet function after mental stress, the increased aggregation by ADP and serotonin after mental stress may be explained by increase in epinephrine and serotonin in plasma after mental stress. The increased levels of serotonin and epinephrine may account for enhanced platelet responses to ADP or serotonin [19]. In addition, rise in plasma catecholamines observed after exposure to stress may damage the vessel wall leading to platelet activation and thrombus formation [20]. Enhanced platelet aggregation after mental stress means that coagulation is enhanced after mental stress. Although the role of serotonin in the initial hemostasis is not fully understood, these results indicate that serotonin in plasma has some role in hemostasis. Physiological role of serotonin may be to facilitate hemostasis by promoting platelet aggregation and by causing local vasoconstriction at sites of injury. It can activate platelets and potentiate the effects of other aggregating agents. When released from platelets it can act as a powerful vasoconstrictor. Serotonin can amplify the effects of catecholamines on the blood vessel smooth muscle. The amplifying action of the amine is more important than its direct constricting effect.

REFERENCES

1. Marsh M. (1981) Normal fibrinolysis, common causes of variation. In: Fibrinolysis. John Wiley & Sons, USA, pp. 92-124.

2. Ogston D. (1983) The influence of physical activity, mental stress and injury on the hemostatic mechanism. In: The physiology of hemostasis. Croom Helm, London, pp. 327-342.

3. Gordon JL, Bowyer DE, Evans DW, Mitchinson MJ (1973) Human platelet reactivity during surgical diagnostic procedures. J. Clin. Pathol. 26: 958-962.

4. Haft JI, Arkel YS (1976) Effect of emotional stress on platelet aggregation in humans. Chest 70: 501-505.

5. Larsson PT, Hjemdahal P, Olsson G, Egberg N, Hornsta G. (1989) Altered platelet function during mental stress and adrenaline infusions in humans: evidence for increased aggregability in vivo as measured by filtragometry. Clin. Sci. 76: 369-376.

6. Levine SP, Twoell BL, Suarez AM, Knieriem LK, Harris MM, George JN (1985) Platelet activation and secretion associated with emotional stress. Circulation 71: 1129-1134.

7. Macht DI (1952) Influence of some drugs and of emotion on blood coagulation. J. Am Med. Assoc. 148: 265-270.

8. Dreifuss F (1956) Coagulation time of the blood, level of blood eosinophils and thrombocytes under emotional stress. J. Psychosom. Res. 1:252-257.

9. Ogston D, McDonald GA, Fullerton HW (1962) The influence of anxiety in tests of blood coagulability and fibrinolytic activity. Lancet ii: 521-523.

10. Macfarlane RG, Biggs R (1946)   Observations of fibrinolysis.   Spontaneous activity associated with surgical operations, trauma, etc.   Lancet ii: 862-864.

11. Jern C, Eriksson E, Tengborn L, Risberg B, Wadenvik H, Jern S (1989) Changes of plasma coagulation and fibrinolysis in response to mental stress. Thromb. Haemost. 62: 767-771.

12. Duke WW (1910)   The relation of blood platelets to hemorrhagic disease. JAMA 55: 1185-1192.

13. Urano T, Sakakibara K, Rydzewski A, Urano S, Takada Y, Takada A. (1990) Relationships between euglobulin clot lysis time and the plasma levels of tissue plasminogen activator and plasminogen activator inhibitor 1.   Thromb. Haemost. 63: 82-86.

14. Takada A, Shizume K, Ozawa T, Takahashi S, Takada Y (1986) Characterization of various antibodies against tissue plasminogen activator using highly sensitive enzyme immunoassay.   Thromb. Res. 42: 63-72.

15. Takada Y, Takada A (1988)   Measurement of the concentration of free plasminogen activator inhibitor (PAI-1) and its complex with tissue plasminogen activator in human plasma.   Thromb. Res. Suppl. VIII: 15-22.

16. Rydzewski A, Takada Y, Takada A (1989) Determination of plasminogen activator inhibitor-1 (PAI-1) in plasma using two different anticoagulants and methods.   Thromb. Res. 55: 285-289.

17. Anderson GM, Young JG, Ohen DJ, Schlicht KR, Patel N. (1981)   Liquid chromatographic determination of serotonin and tryptophan in whole blood and plasma.   Clin. Chem. 5: 775-776.

18. Van der Hoorn FAJ, Boomsma F, Man In't Veld AJ, Schalekamp MADH. (1989)   Determination of catecholamines in human plasma by high performance liquid chromatography.   J. Chromatogr. 487: 17-28.

19. Urano T, Sumiyoshi K, Nakamura M, Mori T, Takada Y, Takada A. (1990) Fluctuation of tPA and PAI-1 antigen levels in plasma: Difference of their fluctuation patterns between male and female.   Thromb. Res. 60: 55-62.

20. Oikawa M, Hsueh AJW (1989)   β-adrenargic agents stimulate tissue plasminogen activator activity and messenger ribonucleic acid levels in cultured rat granulosa cells.   Endocrinology 125: 2550-2557.

21. Santell L, Levin EG (1988)   Cyclic AMP potentiates phorbol ester stimulation of tissue plasminogen activator release and inhibits secretion of plasminogen activator inhibitor-1 from human endothelial cells.   J. Biol. Chem. 263: 16802-16808.

HALOPERIDOL KINETICS IN SENILE DEMENTIA

M. Nishimoto, T. Itaya, Ko. Ohara, K. Miyasato and Ke. Ohara
Department of Psychiatry and Neurology, Hamamatsu University School of Medicine, 3600 Handa-cho, Hamamatsu, 431-31, Japan

INTRODUCTION

The number of patients with dementic disease has increased and psychotic symptoms and behavioral disturbances with dementia are also observed.[1] In these cases, antipsychotic drugs such as haloperidol has been widely used, the dosage generally based on a doctor's clinical experience. Particularly in aged patients, individual deviation in pharmacokinetics becomes large and therapeutic dosage may often induce side effects such as the extrapyramidal syndrome. Thus, in the aged, it is especially important to establish an administration scheme on the basis of pharmacokinetic findings. Haloperidol is the most commonly prescribed antipsychotic drugs. Plasma concentration has been examined for all of the antipsychotic drugs, but haloperidol is considered the ideal agent for such studies because of the drug's relatively simple metabolic pathway.[2] Although, there have been many reports on the plasma concentration and clinical effects of haloperidol, those on effective plasma levels in aged patients are few. In the present study, we compared the plasma levels of haloperidol in the dementic patients above the age of 65 years who were given haloperidol due to psychotic symptoms such as delirium, hallucination, delusion and behavioral disturbances with those in schizophrenic patients under the age of 65.

METHOD

Patient Detail The participants in this study were 31 inpatients being treated in the University Hospital and associated hospital. Sixteen patients (seven men and nine women, 13 vascular dementia and 3 senile dementia of Alzheimer type) were compared with 15 well controlled chronic schizophrenic patients (eight men and seven women). All subjects met DSM-III criteria.[3] The mean age of the patients in the group above 65 was 70.6 years old while that of the patients in the group below 65, 36.3 years old. Mean body weight was 42.6 kg for patients above 65 and 62.3 kg for those below; body weight of patients above 65 was 70 % that of those below. The average daily dose was 1.68 mg for patients above 65 and 11.2 mg for those below.

Study Design Dosage which varied 0.75 mg - 18 mg per day was at the discretion of the doctor in charge. Cerebral vasodilating agents and cerebral metabolic agents were used in combination. Patients who were on other neuroleptics or barbiturates were excluded. Antiparkinson drugs in dosages as low as sufficient for control of adverse effects of haloperidol were allowed as needed. Biperidine, 3 - 10 mg per day, was given to 30 patients, and trihexyphenydyl, 3 mg per day, to 5 patients. Benzodiazepines as sleep inducers were prescribed in cases where patients complained of insomnia. Patients who show abnormality in hematological data, liver or kidney function or ECG were excluded.

After obtaining informal or written consent from family members and patients before the study, haloperidol at doses of 0.75 mg - 6 mg (0.75 - 18 mg per day) was given at 7:30 am. Blood samples were taken at 15, 30, 60, 90 minutes and 2, 3, 6, 11 hours after administration, to determine plasma levels by high performance liquid chromatography (HPLC). Plasma samples separated by a cool centifuge were stored at -60°C until assay. Concentration of haloperidol was determined simultaneously using reverse-phase liquid chromatography with electrochemical detection.[4]

Statistical Analysis Difference between the mean parameter values in patients above 65 and in patients below 65 were examined by unpaired Student's t-test. A difference was not regarded as statistically significant unless the test result had a probability of less than 5%.

RESULTS

The relationship between the plasma levels of haloperidol to daily dose per body weight and age is shown in Fig.1. A weak though significant correlation (r=0.74) was shown, when the groups above and below 65 were combined together, but could not be detected in the group above 65 only.

Fig. 2 shows individual data of patients above 65. While approximately 70% of them showed plasma levels in the range of 3 - 17 ng/ml, this being optimal, the maximum level was about 30 times the minimum, indicating marked deviation among individuals.

Fig. 3 shows the relationship between daily dose per body weight and plasma levels. A significantly high correlation (r=0.89) was confirmed in the group of patients over 65 (black dots). Patients under 65 all appear in the right bottom corner. It indicates that plasma haloperidol calculated from body weight was higher in patients above 65 than in those below. There was no correlation between plasma levels of haloperidol and its daily dose per body weight in the group under 65.

Fig. 4 shows time-course change in plasma haloperidol levels per body weight of single doses. Based on Fig. 4, one could estimate that dose-normalized plasma levels are about 2 times higher in the group above 65 than the standard dose in the group below.

Table 1 shows pharmacokinetic parameters of haloperidol in the group above and below 65. No significant differences in the time lag (Tlag), the time required to reach the maximum plasma level (Tpmax) and absorption coefficient (Ka) could be detected. The elimination half-life ($t_{1/2}$) was significantly longer in patients above 65 than in those below, i.e., the half-life was long and AUC for single dose was about 2 times larger in patients above 65.

Haloperidol was effective in managing target behavioral symptoms in dementic patients. Three dementic patients showed daytime drowsiness and the extrapyramidal symptoms.

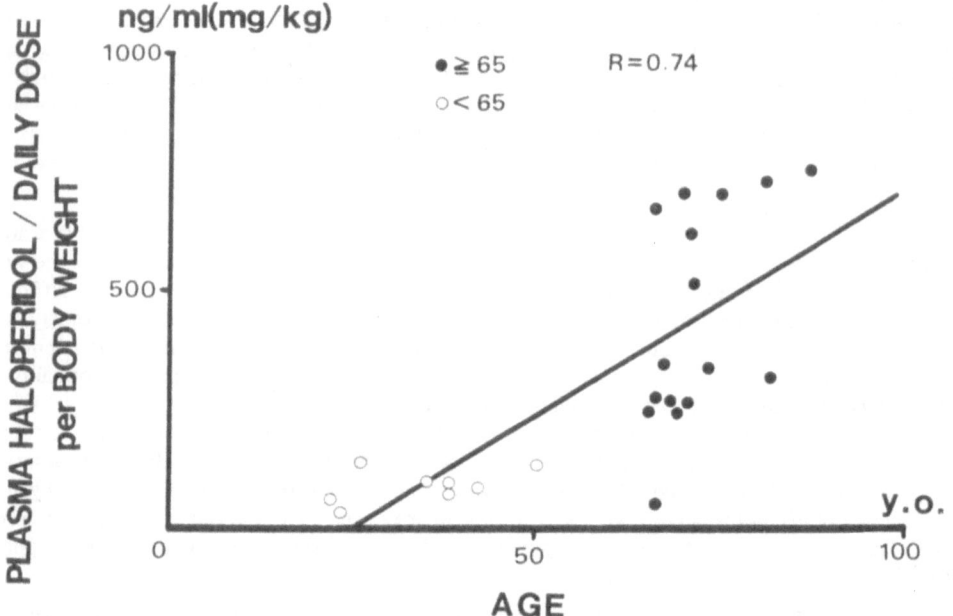

Figure 1: The relationship between plasma levels of haloperidol for daily dose per body weight (kg) and age (year old).

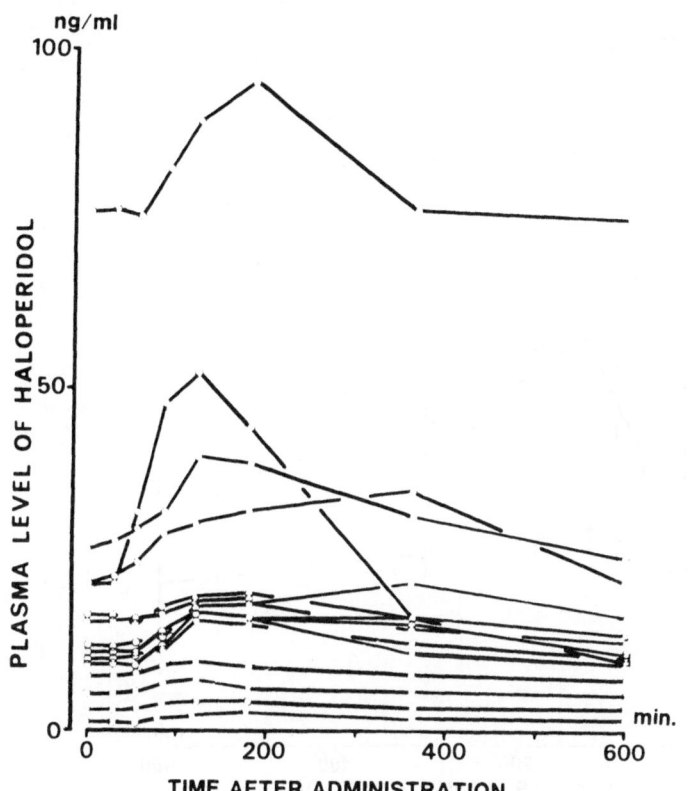

Figure 2: Individual time-course changes in plasma haloperidol levels in patients above 65.

Figure 3: Correlation between plasma levels of haloperidol and daily dose per body weight (kg).

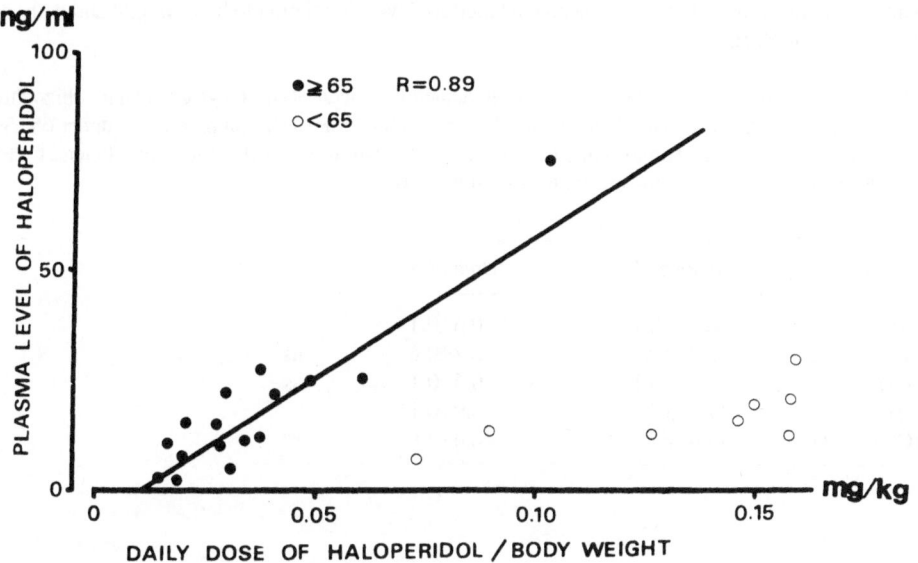

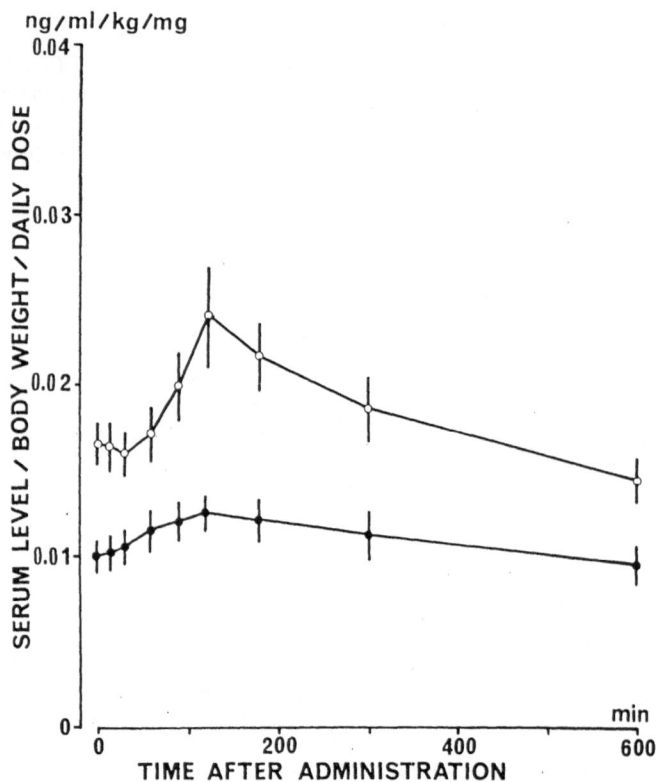

Figure 4: Time-course changes in plasma haloperidol levels in relation to body weight and daily dose. Values are mean ±S.E.

Table 1. The parameters of haloperidol kinetics in dementic patients above 65 years old and schizophrenic patients under 65 years old. Values are mean ±S.E. *:P<0.05, **:P<0.01 compared with under 65. Tlag: The time lag of absorption, Tpmax:Time to reach peak concentration, Ka:absorption coefficient, $t_{1/2}$:Half-life, AUC:Area under the plasma concentration-time curve.

Parameters	Above 65	Under 65	
Tlag(hr)	0.6 ±0.1	0.6±0.1	ns
Tpmax	2.3 ±0.4	2.7±0.6	ns
Ka(/hr)	0.5 ±0.1	0.6±0.1	ns
$t_{1/2}$(hr)	17.3 ±0.1	13.9±0.1	*
AUC(ng/ml.hr)	275.4±67.3	146.0±7.8	**

DISCUSSION

Haloperidol, a butyrophenone neuroleptic agent, is widely used in the treatment of schizophrenic and other psychotic disorders. However, reports on the plasma levels of haloperidol in aged patients, especially in dementic patients with psychotic symptoms, are few. Few studies have attempted to define guidelines for the use of psychotropic medications in the treatment of behavioral symptoms. Most studies have focused on psychogeriatric patients, often excluding those with dementia.[5,6,7,8,9]

Steel et al.[10] conducted a comparative study between haloperidol and thioridazine in the management of behavioral symptoms of 16 patients (mean age, 76) suffering from senile dementia of Alzheimer type. Both drugs were found to be effective in managing target behavioral symptoms, but extrapyramidal side effects were significantly higher with haloperidol. However, they did not determine plasma levels of the drugs.

Neuroleptic agents such as haloperidol readily induce side effects in the senile due to decreased physical functions accompanying aging. The present study is thus significant from this point of view.

The upper and lower limits of the proposed haloperidol therapeutic window are 12 and 2 ng/ml, respectively.[11] Several previous studies indicated that plasma levels of haloperidol vary greatly among individuals. The highest level could be 5 times the lowest level, even when the patients were administered the same dose per body weight.[2] In the present study, 70 % of patients above 65 showed plasma levels of haloperidol from 3 - 17 ng/ml. However, plasma haloperidol levels showed large deviation among individuals. When the results of the schizophrenic patients under 65 and those for patients over 65 were examined together, a significant correlation, though weak, was found between plasma levels per daily dose/ body weight and age but not in the case of the group above 65.

From these results, individual differences in pharmacokinetics, rather than age, appear to be more important. Because a reduction in the hepatic and renal blood flow in geriatric patients and consequent decreases in rates of first-pass effect of the liver as well as a slower rate of clearance may affect plasma haloperidol concentration.[12,13]

A significantly high correlation was observed between daily dose of haloperidol per body weight and plasma levels in patients above 65. The plasma haloperidol level for dosage based on body weight was extremely high in the patients above 65 compared to those below. It is possible that haloperidol may thus readily induce side effects such as extra-pyramidal symptoms in patients above 65. Therefore, the plasma haloperidol level for dosage based on body weight is an important factor for preventing side effects, especially in patients above 65.

In the present study, it was shown that the elimination half-life ($t_{1/2}$) was significantly long in the group above 65, and AUC for a single to be about 2 times larger in the group above 65. From these results, it would appear that when haloperidol is administered to senile patients above 65, dosage should be determined on the basis of a standard set at half of the dose given to a patient under 65 and adjustment made by monitoring clinical effects and plasma levels. This would prevent side effects such as extrapyramidal symptoms and behavioral symptoms, treatment would be based on the basis of optimal plasma level.

REFERENCES

[1] Rabins PV, Mace NL, Lucas MJ (1982) The impact of dementia on the family. JAMA 248: 333-335

[2] Forsman A, Folsch G, Larsson M, Ohman R (1977) On the metabolism of haloperidol in man. Curr. Ther. Res. 21: 606-617

[3] American Psychiatric Association. DSM-III: Diagnostic and Stastical Manual of Mental Disorders. 3rd. Washington, DC: APA, 1980

[4] Korpi ER, Phelps BH, Granger H, Chang W-H, Linnolia M, Meek JL, Wyatt RJ (1983) Simultaneous determination of haloperidol and its reduced metabolite in serum and plasma by isocratic liquid chromatography with electrochemical detection. Clin Chem 29: 624-628

[5] Tewfick GI, Jain VK, Harcup M (1970) Effectiveness of various tranquilizers in the management of senile restlessness. Gerontol Clin 12: 351-359

[6] Tsuang P, Lu LM, Stotsky BA (1971) Haloperidol versus thioridazine for hospitalized psychogeriatric patients: Double-blind study. J Am Geriatr Soc 19: 593-597

[7] Rada R, Kellner R (1976) Thiothixene in treatment of geriatric patients with chronic organic brain syndrome. J Am Geriatr Soc 24: 105-109

[8] Cahn LA, Diesfeldt HFA (1973) The use of neuroleptics in the treatment of dementia in old age. Psych Neurol Neurochir 76: 411-420

[9] Reifler BV, Larson E, Cox G (1982) Treatment results at a multispecialty clinic for the impaired elderly and their families. J Am Geriatr Soc 29: 579-582

[10]Steele C, Lucas MJ, Tune L (1986) Haloperidol versus thioridazine in the treatment of behavioral symptoms in senile dementia of Alzheimer's type: Preliminary findings. J Clin Psychiatry 47: 310-312

[11]Putten TV, Marder SR, Mintz J, Poland RE (1988) Haloperidol plasma levels and clinical response: A therapeutic window relationship. Psychopharmacol Bull 24:172-175

[12]Greenblatt DJ, Sellers EM, Shader RI (1982) Drug disposition in old age. New Engl J Med 306: 1081-1088

[13]Aoba A, Yamaguchi N, Shido M (1986) Difference in the age effect on plasma neuroleptic levels between haloperidol and chlorpromazine in psychiatric patients. In: Clinical and Pharmacological Studies in Psychiatric Disorders. John Libbey and Company Ltd., London: 222-229

HAEMOSTATIC STUDIES IN URAEMIC PATIENTS FOLLOWING DDAVP ADMINISTRATION

M. MYSLIWIEC[1], J. S. MALYSZKO[1], J. MALYSZKO[1], M. PIETRASZEK[2], A. AZZADIN.[2] W. BUCZKO[2]
[1] Nephrology and [2]Pharmacodynamics Departments , Medical School, Bialystok , Poland.

INTRODUCTION

Chronic renal failure is associated with various haemostatic defects. Patients with chronic renal failure have increased bleeding tendency. The laboratory parameter which appears to correlate the best with the risk of haemorrhage is the bleeding time. Current therapy to correct uraemic bleeding includes dialysis intensification, blood and cryoprecipitate transfusions, estrogens or 1- deamino -8 -D -arginine vasopressin (DDAVP, desmopressin). DDAVP is a synthetic vasopressin analogue devoid of vasoconstrictory action which shortens the bleeding time in patients with mild haemophilia [1], von Willebrand disease [1] or platelet function defects [2,3].

In 1982 Watson i Koegh [4] were the first to show that intravenous infusion of DDAVP shortens bleeding time in uraemia. Similar observations have also been reported by others [5,6,7]. In spite of extensive research the precise mechanism by which this drug shortens prolonged bleeding time is still unknown. Mannucci and Ruggeri [8] suggest that DDAVP cause an increase in the plasma concentration of von Willebrand factor , especially its high-molecular-weight multimers. Grant et al [9] have found an increase of noradrenergic system activity. It is possible that both these mechanisms may play a role in the interaction between platelets and the vessel wall. On the other hand there are some data that serotonin (5-hydroxytryptamine) plays an important role in this process [10].

Therefore we have studied some parameters of haemostasis including serotoninergic mechanisms in uraemia and their changes after DDAVP administration.

MATERIALS AND METHODS

33 patients with chronic renal failure (20 females and 13 males, age range 27-66 years) and 9 healthy volunteers (4 females and 5 males, age range 26-35, were studied). They were fully informed of the aims of the study and gave their informed consent. All the uraemic patients were chronically haemodialyzed three times a week for 4-5 hours using cuprophane dialysers. None

of the patients investigated had received blood transfusions for
at least 1.5 months and apart from heparin no drugs which could
influence platelet function were administered for at least 10
days before the study.

All the patients were studied in a double-blind crossover
study. The patient received either placebo (0.9% NaCl) or
desmopressin (Adiuretin SD Spofa, Prague, Czechoslovakia) and then
switched to either placebo or DDAVP after 5 days. Patients were
selected at random to receive DDAVP or placebo.

All the tests were carried out on the day between dialyses.
Patients received neither food nor morning drugs. Between 8.00
and 9.00 after overnight fast and after a rest period of 30 min
in a supine position DDAVP was infused intravenously at a dose
of 0.4 µg/kg b.w. over 20 minutes. Blood samples were taken for
coagulation tests and bleeding time was measured just before and
90 min after DDAVP administration. In all cases the first 0.5 ml
of blood withdrawn from the cannula was discarded. All blood
samples were collected into 0.11 M sodium citrate in 9:1 volume
ratio and platelet poor plasma was obtained by centrifugation at
2500 g for 15 min at 20 ^{o}C(in addition the centrifugation at
4°C was performed for vWFRCof and vWFR:Ag assays). For
measurements of t-PA activity blood was immediately acidified by
adding 1 volume of 0.2 M sodium acetate buffer pH 3.9 to 1
volume of citrated blood and centrifuged without delay. The
plasma samples were aliquotted and used immediately or stored at
- 25°C until assayed.

The bleeding time was measured by the method of Ivy modified
by Mielke [11]. Von Willebrand factor related antigen (vWFR:Ag)
was measured by rocket immunoelectrophoresis according to
Laurell [12] using commercial antisera from Behringwerke,
Marburg, Germany. Factor VIII coagulant activity (VIII:C) was
assayed by one - stage clotting method . Deficient plasma was
obtained from Bio-Merrieux. Ristocetin cofactor of vWF
(vWFRCof)was measured using "von Willebrand Factor assay kit"
(Morris Plains, USA). vWF dependent agglutination using a
standardized platelets fixed with formalin was determined.
FVIII:C, vWFR:Ag, vWFRCof assays were standardized against a
calibrated pool of normal human plasmas. Multimeric state of vWF
was studied according to Bukh et al [13]. Plasma serotonin
(5-HT) concentration was measured according to Aschroft and
Crawford [14]. Platelet serotonin content was assayed using
fluoroscopic method of Drummond and Gordon [15]. ^{14}C serotonin
release from plateletes was determined by method of Lingjaerde
[16] and serotonin uptake was measured as described by Gordon
and Olverman [17]. Fibrinolysis was assayed by measuring the
euglobulin clot lysis time (ECLT) [18]. Tissue plasminogen
activator (t-PA) and tissue plasminogen activator inhibitor
(PAI) activities were determined by measuring the amidolytic
activity of plasmin on chromogenic substrate S-2251 (Kabi
Vitrum, Sweden) as described by Chmielewska and Wiman [19].

Values given are means ± SD. Statistical analysis was

performed by Student's t-test for paired data with p < 0.05
considered significant.

RESULTS

Table I.Results of the studies of haemostasis in control and
uraemic patients before and after DDAVP administration

	control	uraemics before DDAVP	after DDAVP
Bleeding time (min)	6.2±1.75	20.6±8	11.9±8.3 ***
VIII:C (%)	101.3±18.4	247±84.5	297.7±170
vWF:Ag (%)	93.7±22.6	239.1±94	473.1±243***
vWFRCof (%)	101.6±30.2	231±162	347.3±178***
ECLT (min)	154.1±31.3	257.8±97.9	163.9±96***
t-PA (IU/ml)	1.93±0.84	1.87±0.75	3.4±0.9***
PAI (AU/ml)	8.38±1.81	7.77±2.05	4.41±1.1**
Plasma 5-HT (ng/ml)	2.6±0.9	2.5±1.1	4.2±1.0*
Platelet 5-HT (ng/10^{9}plat)	397±70	541±205	409±88*

* p < 0.02 versus pre-treatment value
** p < 0.01
*** p < 0.005

Bleeding time was shortened significantly from 20.6 to 11.9
min after DDAVP administration as well as ECLT from 257.8 to
163.9 min. A parametric negative correlation was obtained
between bleeding time and ECLT (r=-0.43) A significant rise in
vWF:Ag level was found while VIII:C activity increased from 247
to 297.7%.It was not the significant change. An increase in vWF
multimers pattern after DDAVP administration was observed.t-PA
activity was raised significantly from 1.87 to 3.4 IU/ml and PAI
activity fell following DDAVP administration from 7.77 to 4.41
AU/ml .Platelet serotonin content was significantly reduced from
541 to 409 ng/10^{9}plat and plasma serotonin concentration raised
from 2.5 to 4.2 ng/ml .A parametric correlation was found
between the rise in t-PA activity and the decrease in platelet
serotonin content (r=0.77) as well as between the decrease in

PAI activity and the fall in platelet serotonin content
(r=-0.57) . It was observed that DDAVP inhibited ^{14}C serotonin
uptake in a dose-dependent manner whereas ^{14}C serotonin release
was increased. After 2 hours of platelet incubation with DDAVP
15% of ^{14}C serotonin was released from platelets.

Table II.Studies of ^{14}C serotonin uptake and release after
incubation of normal and uraemic platelets with DDAVP at a
concentration of 4000 pg/ml.

	normal platelets before DDAVP	after DDAVP	uraemic platelets before DDAVP	after DDAVP
^{14}C serotonin uptake (pmol/10^{8}plat)	150±14	105±8.8**	164±12	100±9**
^{14}C serotonin release (%)	100	87*	100	85*

* p < 0.05 versus pre-treatment value
** p < 0.01

DISCUSSION

   Bleeding time apparently correlates the best with bleeding
tendency.Its prolongation in uraemics has been a continuous
finding.We also observed an increase in the compounds of factor
VIII in uraemics and prolongation of the ECLT without changes in
t-PA and PAI activities.Plasma serotonin level was normal and
platelet serotonin content was found to be increased in
uraemia.No changes in ^{14}C serotonin uptake and its release by
uraemic and control platelets in vitro were found.DDAVP
shortened the bleeding time in uraemics and induced further
increase in factor VIII compounds as well as increased vWF
multimers in blood.It shortened the ECLT and markedly increased
plasma t-PA activity at the same time decreasing plasma PAI
activity.Platelet serotonin content was decreased whereas plasma
serotonin concentration was markedly raised. ^{14}C serotonin uptake
by uraemic platelet in vitro was decreased and ^{14}C serotonin
release was raised.
   The mechanism of a haemostatic action of DDAVP in uraemic has
been still obscure.It has been suggested that DDAVP acted by
releasing vWF particularly its high-molecular-weight multimers
and/or proteins from cellular storages other than the platelets
[8].However it was not confirmed by others [20].
   On the other hand DDAVP causes a marked increase in
circulating fibrinolytic activity mainly due to a rapid release
of t-PA into the blood stream.According to Moffat et al [21] and
Mannucci et al [22] desmopressin does not act directly on
cultured endothelial cells or isolated perfused vessels.

167

Therefore the action of this drug on the plasma fibrinolytic activity may be indirect. Brommer and co-workers [23] found that the fibrinolytic response to DDAVP could not be inhibited by the betaadrenoreceptor blockade. The possibility of prostaglandin mechanism of action of DDAVP on fibrinolysis was also excluded by Brommer et al [23]. There has been no evidence for increased platelet activation by desmopressin as tromboxane B_2 , betathromboglobulin and platelet factor 4 levels also remain unchanged [24].

Our data indicate that DDAVP causes a fall in platelet serotonin content [25]. It may be due to a release of serotonin from platelets and its impaired uptake. Serotonin, a potent vasoconstrictor, can activate platelet potentiating the effect of other aggregating agents. But according to Yang et al [26] it is unlikely that the beneficial effect of desmopressin in patients with primary platelet defects is related to a direct stimulatory effect on platelets. We found a significant dose-dependent DDAVP induced rise in platelet serotonin uptake although there was a paper showing contrary data [27].

We therefore postulate that a messenger of the haemostatic action of DDAVP could be serotonin which is released from platelets to plasma by this drug. A concomitant increase in t-PA activity seems to be a counteracting mechanism on the shortening of bleeding time by desmopressin. Strong correlation between the release of t-PA and the fall of platelet serotonin content after DDAVP administration suggests a possible role of this amine also as a mediator of action of this drug on fibrinolysis.

REFERENCES

1. Mannucci PM. Ruggeri ZM. Pareti FJ. Capitanio A (1977) Lancet 1: 869-872

2. Kobrinsky NL. Israels ED. Gerrard JM. Cheang MS. Watson CM. Bishop AJ. Schroeder ML (1984) Lancet 1: 1145-1148

3. Schulman S. Johnsson H. Edberg N. Blomback N (1987) Thromb Res 45: 165-174

4. Watson AJS. Koegh JAB (1982) Nephron 4: 260-261

5. Mannucci PM. Remuzzi G. Pusineri F. Lombardi R. Valsecchi C. Mecca G. Zimmerman TS (1983) N Engl J Med 308: 8-12

6. Rydzewski A. Rowinski M. Mysliwiec M (1986) Folia Haematol (Leipzig) 113: 823-830

7. Malyszko J. Pietraszek M. Buczko W. Mysliwiec M (1990) Folia Haematol (Leipzig) 117: 319-324

8. Ruggeri ZM. Mannucci PM. Lombardi R. Federici AB. Zimmerman TS (1981) Blood 59: 1272-12

9. Grant MB. Guay C. Lottenberg R (1988) Thromb Haemostas 59: 269-272

10. DeClerk T. Somers Y. van Gorp L (1984) Agents Action 15: 627-635

11. Mielke CH. Kaneshiro JA. Maher JA. Wessler JM. Rapaport SI (1969) Blood 34: 204-215

12. Laurell CB (1966) Anal Biochem 15: 45-52

13. Bukh A. Ingerslev J. Stenbjerg S. Hundahl-Moller NP (1986) Thromb Res 43: 579-585

14. Achroft G. Crawford TB (1964) Clin Chim Acta 9: 364-369

15. Drummond AH. Gordon JL (1974) Thromb Diathes haemorrh (Stuttgart) 31: 366

16. Lingjaerde D (1979) Acta Pharmacol Toxicol 45: 394

17. Gordon JL. Olverman HJ (1978) Brit J Pharmacol 62: 219

18. Kowarzyk H. Buluk K (1950) Post Hig Med Dosw 2: 1-76

19. Chmielewska JN. Wiman B (1985) Clin Chem 32: 482-485

20. DiMichele DM. Hatheway WmE (1990) Am J Haematol 33: 39-45

21. Moffat EH. Giddings JC. Bloom AL (1984) Brit J Haematol 57: 651-652

22. Mannucci PM. Canciani MZ. Roto L. Donovani BS (1979) Brit J Haematol 43: 283-293

23. Brommer EJP. Derkx FHM. Barret-Bergshoeff MM. Schalekamp MADH (1984) Thromb Haemostas 51: 42-44

24. D'Angelo A. Capitanio A. Smith JB. Vasecchi C. Mannucci PM (1983) Thromb Haemostas 49: 64

25. Malyszko J. Pietraszek M. Azzadin A. Buczko W. Mysliwiec M (1990) Thromb Haemostas 61: 637

26. Yang X. Disa J. Rao AK (1990) Thromb Res 59: 809-810

27. Soslau G. Schwartz AB. Putatunda B. Conroy JD. Parker J. Abel RF. Brodsky I (1990) Am J Med Sciences 299; 372-379

BIOLOGICAL RESPONSES AT ARTIFICIAL SURFACES AND RECENT PROGRESS IN EXTRACORPOREAL CIRCULATION

TAKEHISA MATSUDA

Dept. of Bioengineering, National Cardiovascular Center Research Institute, 5-7-1 Fujishirodai, Suita, Osaka, Japan

INTRODUCTION

The activation of body defence mechanisms during extracorporeal circulation often causes local and systemic adverse effects on the body. The former is exemplified as thrombus formation at the blood-material interfaces, and the latter as peripheral circulation failure associated with microemboli and granulocyte aggregation, resulting in symptoms such as systemic hypertension and peripheral hypotension. Understanding the biological responses at the blood-material interface of extracorporeal devices, such as a hemodialysis, plasmapheresis and artificial oxygenator, is very important for upgrading the biocompatibility of a device under development. The body defence mechanisms associated with thrombus formation and immunological alterations include many biological systems, including the coagulation, complement and cellular systems. As schematically shown in Fig. 1, the multiple activations of these biological systems occur as blood comes in contact with a foreign surface. The characteristic feature of body defence mechanisms leading to thrombus formation is that, although they are independently activated at blood-material surfaces, there exists a positive feed-back mechanism in which an activated form also activates other biological systems. For instance, the activated form (Factor XIIa) of the coagulation system can activate the third component of the complement system, and the activated complement factor (C5a) is a potent aggregation activator of granulocytes. Besides participation in thrombus formation, immunological alterations are involved in humoral and cellular activation. Therefore,

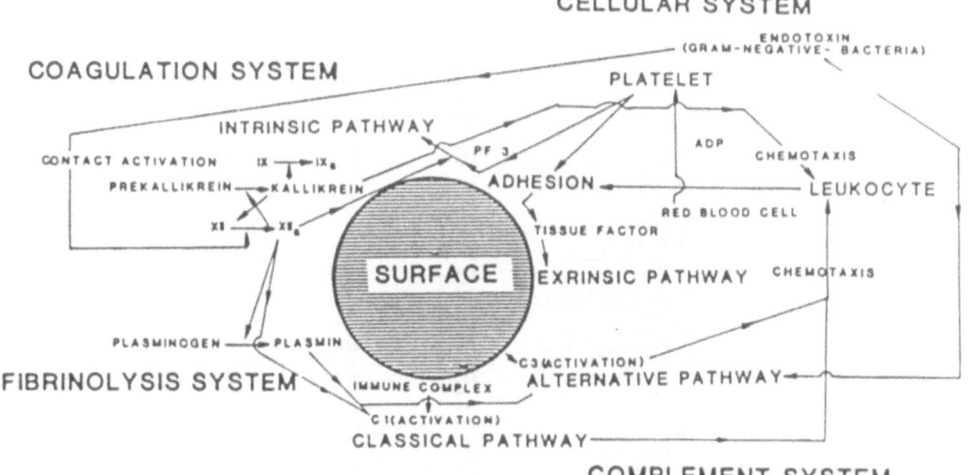

Fig.1. Multi-interacting biological systems leading to thrombus formation on polymer surfaces

the understanding of molecular events at a blood-material interface is of particular importance for the logical surface design of blood-contacting material, particularly that used in extracorporeal devices. This paper summarizes our observations over years on blood-material interactions, focusing especially on the molecular events in the initiation of biological activation on polymeric surfaces with a wide spectrum of surface properties. An attempt is made to establish the 'surface property-biological response' relationship. Finally, a few examples for providing the reliability of blood compatibility were demonstrated via surface design and pharmacological approach.

Coagulation System

The coagulation system is composed of at least ten plasma proteins and proceeds via cascade reactions by either the intrinsic or extrinsic pathway, resulting in fibrin formation at the end stage. The activation of the coagulation system on non-physiological surfaces is initiated by the intrinsic pathway. The initiation reaction is called the contact phase activation, which involves three coagulation factors (Factor XII; FXII, high-molecular weight kininogen; HMWK, and prekallikrein; PK) The formation of the trimolecular complex on a surface is the essential requirement for activation. Upon activation for the trimolecular complex, limited proteolysis generates the activated form of the coagulation factors. These activate the coagulation factor following in the coagulation cascade (Factor XI). The quantitative measurement of the amount of kallikrein, which is converted from PK upon activation, can be easily determined using the fluorescent synthetic substrate Z-Phe-Arg-MCA. This fluorogenic assay method was found to be very specific and sensitive to quantitative measurements of kallikrein. At an enhanced surface-to-volume ratio using polymer coated glass beads, this method allows the quantitative determination of which polymer is the most potent activator and which is essentially inert in the coagulation system. The general observations obtained are summarized below (Table 1).

polymers

Polymer	Contact activation*	
	Reconstituted system	Plasma system
1. PDMS	0.00	19
2. PST	0.03	25
3. Poly(butyl methacrylate)	0.07	38
4. Poly(lauryl methacrylate)	0.03	18
5. PVC	0.13	41
6. PET	0.15	97
7. PMMA	0.18	75
8. Polycarbonate	0.30	64
9. Poly(caprolactone)	0.41	59
10. Nylon-6	0.74	81
11. Nylon-66	0.39	64
12. Polyurethane (MDI/BD)	0.38	73
13. PEO	0.06	27
14. Cellulose triacetate	0.24	95
15. Cellulose acetate (CAc)	0.24	140
16. Hydrolysed CAc (cellulose)	0.09	41
17. Poly(vinyl acetate)	0.67	198
18. PVA	0.04	9
19. PVA (heat treated)	0.04	15
20. EVAL 33	0.09	29
21. EVAL 33 (heat treated)	0.11	25
22. Poly(vinyl phenol)	0.06	30
23. PMMA-SO₃H^b	0.71	350
24. PAN-SO₃H^c	0.99	522
25. Glass	0.78	380

*After subtraction from the background level (nmol/ml/min)
^bPoly(2-acrylamido-2-methyl-1-propanesulphonic acid-CO-methyl methacrylate); MMA:95%
^cPoly(2-acrylamido-2-methyl-1-propanesulphonic acid-CO-acrylonitrile); PAN:95%

Table 1. Degree of contact activation of coagulation system on various polymers

1. The most potent activators are found to be negatively charged functional group-bearing surfaces, as shown in Table 1. The best example is surface sulphonated polystyrene. Increasing the surface density of sulphonate groups drastically enhances the activation. The incorporation of carboxyl groups on polymer surfaces also effectively activates the contact phase.
2. Hydrophobic surfaces such as polystyrene and poly (dimethyl siloxane) are essentially inert. Both cationic and hydroxyl group-bearing surfaces inhibit formation of the trimolecular complex resulting in no activation.
3. Among polar surfaces, electron-donating surfaces typified as poly (vinyl acetate) are more active than electron-accepting surfaces. These are classified as weak to mild activators.
4. Electrostatic interaction is required for the adsorption process leading to the formation of the trimolecular complex. Polystyrene surfaces adsorb these coagulation factors at high surface-to-volume ratio but essentially no activation results. This may indicate that a favourable configuration and a conformational change on polymer surfaces would be required for the activation.

Complement System

The complement system has also been found to participate in thrombus formation. Similar to the coagulation system, the complement system also consists of twenty different proteins, which are sequentially activated. Activation of the complement system is by either the classical pathway or the alternative pathway. The former pathway is initiated with the first component of complement factor (C1) and the latter is initiated with the third component C3. Irrespective of pathways, the fragments of activated complement factors, C3a and C5a, are generated upon activation. These are called anaphylatoxins and result in adverse effects on various cellular systems. Table 2 gives the complement activation

Polymer	Generated anaphylotoxin (ng/ml)			Complement titer	
	C3a Alternative	C3a Classical	C4a	ACH_{50}(%)	CH_{50}(%)
1. PDMS	2430	−100	180	79.1	104.6
2. PST	1310	790	450	83.0	91.8
3. Poly(butyl methacrylate)	1070	120	840	87.6	86.3
4. Poly(lauryl methacrylate)	670	830	540	88.8	86.8
5. PVC	2570	1280	2050	80.3	82.5
6. PET	1220	1030	1400	84.4	86.2
7. PMMA	1650	920	550	88.1	87.8
8. Polycarbonate	−180	380	680	94.8	88.1
9. Poly(caprolactone)	4570	1580	1950	81.8	80.0
10. Nylon-6	1820	2630	3200	88.2	84.7
11. Nylon-66	1200	2750	2400	95.4	85.9
12. Polyurethane (MDI/BD)	3790	1260	1200	81.8	90.1
13. PEO	−180	210	450	102	93.0
14. Cellulose triacetate	5770	1380	1580	84.4	79.6
15. Cellulose acetate (CAc)	10970	1480	1200	70.4	76.4
16. Hydrolysed CAc (cellulose)	10070	1380	1220	—	—
17. Poly(vinyl acetate)	11970	4080	5950	53.1	75.3
18. PVA	13070	1280	2000	87.3	91.8
19. PVA (heat treated)	15670	1880	3050	81.4	72.9
20. EVAL 33	3820	8230	14140	91.1	78.8
21. EVAL 33 (heat treated)	—	—	—	—	—
22. Poly(vinyl phenol)	250	−350	900	88.9	89.0
23. PMMA-SO₃H[b]	2850	1750	1900	92.0	84.6
24. PAN-SO₃H[c]	3410	2040	2450	91.5	79.6
25. Glass	650	30	480	91.5	100

[a]After subtraction from the background level
[b]Poly(2-acrylamido-2-methyl-1-propanesulphonic acid-CO-methyl methacrylate); MMA:95%
[c]Poly(2-acrylamido-2-methyl-1-propanesulphonic acid-CO-acrylonitrile); PAN:95%

Table 2. Degree of complement activation on various polymers

characteristics of a wide variety of polymer surfaces. The surface reactivity with regard to complement activation was assessed by conventional haemolytic titre and RIA-methods. RIA-method was employed to measure anaphylatoxins with and without the presence of calcium ions, thus differentiating the contribution of the classical pathway from overall activation. The general trend shows :

1. The most inert surfaces are found to be hydrophobic.
2. The hydroxyl group-bearing surfaces are the most potent activators as typified by poly (vinyl alcohol) PVA and ethylene-vinyl alcohol copolymer EVAL.
3. Some polar surfaces preferably activate by the alternative pathway, but EVAL preferentially activates by the classical pathway.

In order to define the initiation reaction of the complement activation by both pathways at molecular levels, experiments using reconstituted systems of purified complement factors were designed. Physiologically, the initiation reaction in the classical pathway involves three proteins of the first complement component (factors Clq, Clr and Cls), and is similar to the contact phase of the coagulation system (Fig. 2). Physiologically, activation of Cl is initiated with the pentamolecular complex of the Cl component formed on the Fc region of the immune complex. The reconstituted system showed that the maximum activation rate was observed at the compositional ratio of Clq:Clr:Cls of 1:2:2, which is identical to that found in physiologically activated states. In the absence of IgG, potent complement activating surfaces such as PVA and EVAL activate the Cl complex in serum to the same extent as thermally aggregated IgG, which has been frequently used as a model activator in the classical pathway. In surfaces with preadsorbed IgG, EVAL drastically reduces its activating power, whereas little change in the activation rate on PVA was observed irrespective of the presence or absence of preadsorbed IgG.

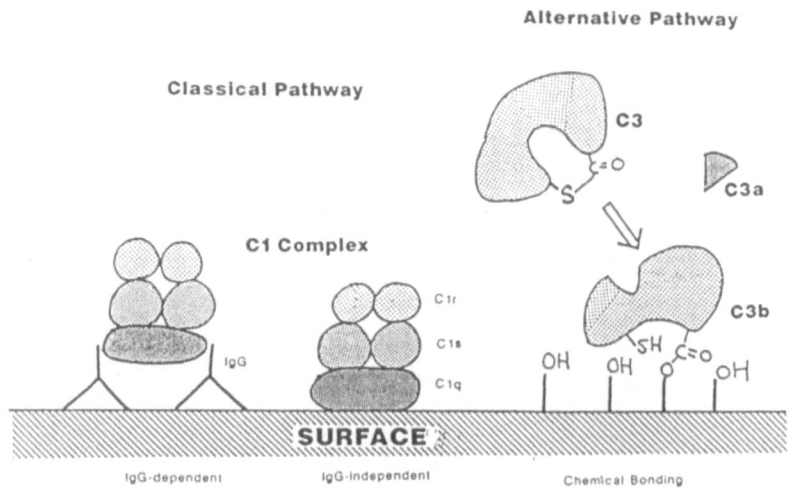

Fig.2. Mechanisms of complement activation on polymer surfaces

These results indicate that there are two mechanisms for Cl activation in the classical pathway. One is by direct complex formation of Cl on surfaces, and the other by complex formation on preadsorbed IgG. In the latter the Fc region of preadsorbed IgG, which is orientated away from the surface, could facilitate the absorption of Clq resulting in Cl complex formation, as found in physiological states.

The alternative pathway is initiated with the activation of C3. Why the hydroxyl group-bearing surfaces significantly activate the complement system has been explained by the following hypothesis : C3b, the activated fragment of C3, binds chemically to the surface hydroxyl groups. In order to verify this hypothesis, radioisotope-labelled purified C3 was activated by various non-physiological and physiological agents in the presence of PVA and cellulose films. The radioactivity after vigorous washing with various high-ionic-strength aqueous solutions containing non-ionic and ionic surfactants (Triton 100 and sodium dodecyl sulphate) remains unchanged. On the other hand, treatment in alkaline aqueous solution results in a sharp decrease in the remaining radioactivity. In addition, treatment with a fluorescent dye (DACM), which selectively reacts with the SH group, showed fluorescence at the surface which was observed using fluorescence microscopy. These observations indicate that C3b definitely binds chemically to the surface hydroxyl groups to form an ester linkage. The localization of chemically fixed C3b on the surface provides a locus for multiple activation and subsequent amplification for the subsequent steps of the alternative pathway.

Heat treatment of PVA results in an enhancement of surface crystalisation. The enhanced activation found on higher temperature-treated PVA shows that the local organization of hydroxyl groups accelerates activation (Fig. 3). This indicates that, in addition to the type of functional groups, the surface density and rigid organized layer also determine the activating power. This, in turn, indicates that interfacial structural design should be incorporated for aquisition of biocompatibility.

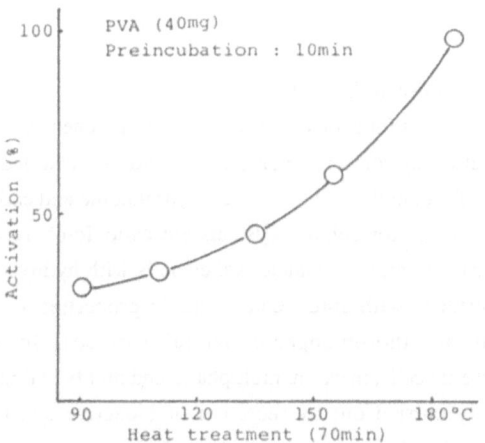

Fig.3. Effect of heat treatment of PVA on Cl activation

Cellular Adhesion

The cellular system containing platelets and leukocytes is the most potent biological system leading to thrombosis. The generally accepted understanding is that adhesion of these cellular systems is greatly promoted on surfaces with absorbed fibrinogen (Fb) and fibronectin (Fn). These proteins, known as adhesive proteins, promote the cellular adhesion of various types of anchorage-depending cells such as fibroblasts. Recently, the minimal amino acid sequence common to the so-called adhesive proteins (Fb, Fn, von Willebrand factor) was identified as arginine-glycine-aspartic acid-serine (RGDS). This implies that these cells have membrane receptors which molecularly recognize the adhesive site of proteins. It is of primary interest to clarify whether the adhesion of blood cells on an adhesive protein-preadsorbed surface is also controlled by the RGDS ligand-receptor interaction. It is evident that the fluorescent-labelled RGDS binds to both cells, as this was clearly demonstrated under fluorescence microscopy. When the synthetic tetrapeptide, RGDS, is premixed with platelets and leukocytes, their adhesion is drastically reduced in a dose-dependent manner. This indicates that RGDS may be an antagonist for the membrane receptor. These results strongly suggest that the adhesion of platelets and leukocytes is mainly controlled via the RGDS ligand-receptor interaction. Therefore, the adsorption of these plasma adhesive proteins on polymer surfaces should be avoided to minimized. In general, hydrophobic surfaces enhance the adsorption of these adhesive proteins. Further evidence of the participation of the RGDS ligand-receptor interaction is the inhibition by RGDS of intracellular calcium, mobilisation of platelets by collagen. Collagen also has the RGDS sequence as an adhesive site. This was assessed by the calcium-sensitive

5protein (aequorin) and probe (Fura-2), which were incorporated into the platelet cytoplasm. This prompted us to use RGDS peptide and its analogues for platelet-preservation during extra-corporeal circulation as mentioned later. Besides biospecific interactions in cellular adhesion as mentioned above,

cationic charged surfaces strongly promote the adhesion of cells, due to a non-specific mechanism driven by electrostatic interaction.

'Surface Property-Biological Response' Relationship

The overall biological response as a function of surface property is schematically summarized in Fig.4. Inert surfaces for the coagulation system are hydrophobic, cationic-charged and hydroxyl-bearing, whereas the most inert surfaces for complement system are hydrophobic and cationic. On the other hand, the most potent adhesion promoters for cellular systems are cationic-charged surfaces followed by hydrophobic surfaces. The least activation of platelet adhesion is with hydroxyl-bearing surfaces. This indicates that none of the surfaces with single-characteristic properties satisfies biocompatibility. Endothelial cells are ideal non-thorombogenic natural surfaces. In terms of the surface structure/composition of endothelial cells, these are multiphasic and highly hydrated. The inherent built-in non-thrombogenicity is mainly carried out by secretion of bioactive substances and metabolites. Currently, various approaches aimed at molecular design of blood-compatible materials have been conducted. An attempt to classify these approaches along multiple functions and properties of endothelial cells is presented in Fig.5. The approach of molecularly designing biocompatible polymers of choice depends on the device under consideration.

Recent Progress on Extracorporeal Circulation

Two examples of extracorporeal devices with minimized activation of body defence mechanism will be presented here. One is a hemodialyzer developed by Prof. Ikada (Kyoto University) and manufactured by Asahi Medical Inc., Japan. As pointed out, hydroxyl group-bearing surfaces such as cellulose\ is the most potent activator for complement system, which often causes leukopenia, hyperoxemia, anaphylatoxin shock, all of them are induced by generated anaphylatoxin C3a and C5a. Therefore, logical surface design was placed on chemical modification of surface hydroxyl group. Poly (ethylene glycol) was grafted via isocyanate-treated hydroxyl group on cellulose hollow fibers. This method could have dual characteristics. One is a minimized density of surface hydroxyl group, reducing complement activation. Second is water-swellable interfacial region created with grafted poly (ethylene glycol) which could reduce the cellular adhesion and protein adsorption. The in vitro and clinical results found agreed with performances predicted along with the working hypothesis mentioned above. The other example was demonstrated with a novel heparin-imgregrated device modification technology depeveloped by Carmeda. AB, Sweden, specially suited for artificial oxygenator. The coupling method of heparin on surfaces was unique, in which oxidized end group of heparin was attached to aminated on surface. This end-attachment method was found to be the most effective in preserving heparin activity as compared with conventional heparin immobilized methods.The Carmed system was proven to impart excellent antithrombogenicity, which was achieved under non-heparinization during extracorporeal circulation such as artificial oxygenator. These approaches were based on physico-chemical and biological considerations on surface biological activity. Along with device's surface modifications, pharmacological alterations of blood is alternative for ensuring to minimize blood-material interaction. As shown in the 'Cellular Response' section, our study showed that the administration of tetrapeptide RGDS effectively blocked the activity of

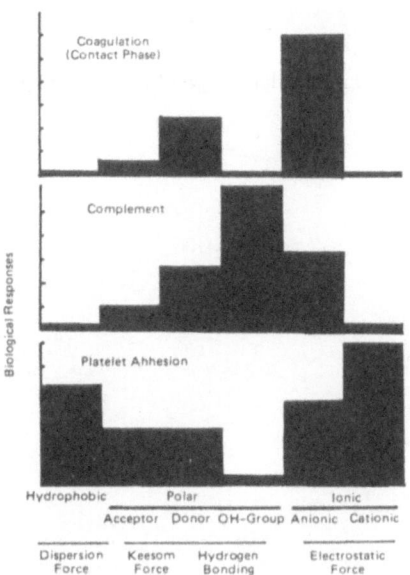

Fig.4. General features of surface blood reactivities

176

cell adhesion during extracorporeal circulation. The short-life of the activity (less than ten minutes) was found to be best-suited for platelet-preservation during artificial oxygenation, because platelet aggregability is recovered shortly after the cession of administration upon weaning from extracorporeal circulation. Thus, the combination of logical surface design and phamacological design based on biological specificity allows to maximize the in vivo performance in advanced medical technology.

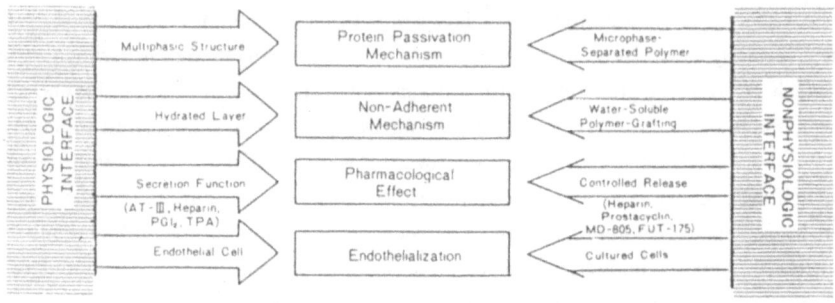

Fig.5. Conceptualised molecular design of blood compatible polymers

ANTITHROMBOGENIC BIOMATERIALS. POLY(VINYL CHLORIDE) WITH HEPARIN- AND/OR PGI$_2$-IMMOBILIZED IN HYDROGELS AND THROMBOMODULIN-IMMOBILIZED BEADS.

IKURO MARUYAMA[1], MITSURU AKASHI[2], & TOSHIHIDE OKADOME[1]
[1]IIIrd Department of Internal Medicine, Kagoshima University School of Medicine, [2]Department of Applied Chemistry Faculty of Engineering , Kagoshima University, Kagoshima Japan.

INTRODUCTION

The luminal surface of blood vessels is covered with endothelium which possesses various antithrombogenic functions and maintain blood fluidity. The anti-thrombogenic functions include the production of PGI$_2$, the secretion of tissue plasminogen activator, and the production of heparin-like proteoglycans and thrombomodulin(TM)(1). PGI$_2$ inhibits platelet functions and causes blood vessel dilatation(2). Tissue plasminogen activator activates plasminogen to plasmin. Heparin-like proteoglycans activate antithrombin III from a progressive inhibitor to an immediate-type inhibitor, which inhibits activated factor Xa and thrombin(3). TM is an endothelial associated glycoprotein that convert thrombin from a procoagulant protease to an anticoagulant(4,5).

In blood compatible polymer materials, it is required to inhibit both platelet adhesion/aggregation and blood coagulation on the polymer surface like as the endothelial-surface. Several antithrombogenic polymer materials have been described including biologically active substance such as heparin (6,10), urokinase, and prostaglandins(8, 9, 10, 11).

Since poly(vinyl chloride)(PVC) has good mechanical characteristics for biomedical use, it is currently one of the most useful biomaterials(6). However, the blood compatibility of PVC is not satisfactory for medical usage. Surface modified and heparinized PVC's have been prepared to achieve antithrombogenicity and being used for blood bags and catheters.

In this paper, we prepared the immobilization of biologically active heparin and/or PGI$_2$ on PVC and evaluated its antithrombogenic activity by measuring inhibiting effect of platelet aggregability and blood clotting. We also attempted to prepare recombinant TM(rTM)-immobilized glass beads and evaluated its antithrombogenic effect.

MATERIALS AND METHODS

Polyacrylamide Gels with heparin or PGI$_2$ The polyacrylamide gels were prepared and immobilized with heparin and/or PGI$_2$ as described previously(6). Human rTM, prepared as previously described(7,13), was kindly supplied by Asahi Chemicals Co. and immobilized on glass beads. Human plasma and platelet rich plasma(PRP) were obtained by centrifugation of citrated normal plasma. All other chemicals were obtained Sigma Co.(St. Louis, Mo) and reagent grade.
Measurement Heparin release from PVC was determined by the method of previously described(6).

Platelet aggregation was measured by aggregometer(Syenco) using collagen(2 μg/ml, Niko Bioscience, Inc.) as an inducer of aggregation(12) in the PVC-treated cuvettes with or without heparin and/or PGI$_2$.

Whole blood clotting time and (activated) partial thromboplastin time(aPTT or PTT) were measured in PVC coated cups with or without

Key words: antithrombogenicity, heparin, PGI$_2$, thrombomodulin, endothelium

heparin by conventional methods using Fibrometer(BBL).

TM was immobilized on the surface of glass using the carboxyl termi-
nate of the molecule. The condensation between amino group of N-(2-
aminoethyl)-3-aminopropyl substituted glass and carboxyl group of TM
using water soluble carbodiimide gave TM-immobilized glass in a 0.2 M
phosphate ·buffer solution. To evaluate the antithrombogenic activity
of the TM-immobilized glass, we used partial thromboplastin time
without kaolin, because the glass in the sample will be act as an
artificial surface. The effect of the TM-immobilized glass on platelet
aggregation, we measured thrombin-induced platelet aggregation with
the presence of TM-immobilized glass or control glass in PRP.

RESULTS AND DISCUSSION
In order to estimate the release of heparin from the PVC matrix into
blood , the amount and the rate of the release of heparin from the PVC
matrix were estimated at 37°C in physiological saline. Too much immobilized heparin in PVC(sample No.3 in Fig.1) is not well sustained ,
however, suitable amounts of heparin(sample No.1 and 2) were effectively immobilized and released slowly from the PVC matrix.

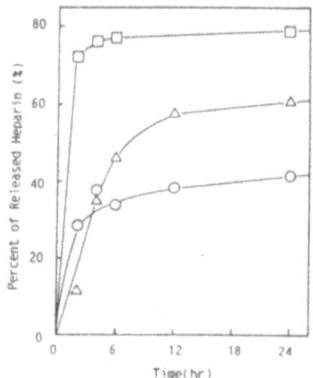

Fig.1. Release of heparin from PVC with heparin-immobilized hydrogels
at 37 C in a physiological saline. (O)Sample No.1,
(△)Sample No.2, (□)Sample No.3

Thus hydrogel is considered as a reservoir for the hydrophilic
biologically active molecules. As the heparin comes out of the reservoir into the hydrophobic PVC matrix it moves quickly through the
hydrophobic domain.
It has been reported that releasing rate of heparin is necessary at
least 4×10^{-5} mg/cm^2/min to prevent formation of thrombus. In this
system heparin-releasing rates in sample #1 and #3 were lower than
the guideline(data not shown). Since we observed in scanning electron
micrograph that a part of PVC-surface was covered by hydrogel, it will
be expected that antithrombogenicity may be achieved in lower heparin
release rate than the guideline. To confirm the idea, we evaluated the
antithrombogenicity of PVC by measuring whole blood clotting time in
the PVC coated cups with or without heparin. The results were summarized in Table 1.

Whole blood clotting time of PVC with heparin- and/or PGI₂-immobilized hydrogels.[a]

Run No.	Hydrogels	Heparin (μg)	Prostaglandin (μg)	Whole blood clotting time (min)		
				0 h	22 h	46 h
10	Sample No. 4	19.1	none	>30	>45	>45
11	Sample No. 9	none	1.04	9.5	7.5	9
12[b]	Sample No. 4 + Sample No. 9	23.3	0.26	>30	>45	>45
13[c]	Sample No. 4 + Sample No. 9	6.76	0.30	>30	>45	>45
14	Sample No. 8	none	none	10	12	14

[a]Hydrogel/(Hydrogel + PVC) = 10 wt%
[b]Sample No. 4/Sample No. 9 = 4/1 (wt/wt).
[c]Sample No. 4/Sample No. 9 = 1/1 (wt/wt).

In the cup coated with PVC containing hydrogels without heparin(sample No. 8), blood clotting occurred immediately.
However the PVC's containing heparin immobilized gels prevented clotting for 24 h.

It has been described that the biomaterials with immobilized prostaglandins show excellent antithrombogenic activity by inhibiting platelet aggregation. Since natural form of prostaglandins are too unstable to use, biologically active and stable prostacyclin analogue beraprost was used in this study.

As shown in Figure 2, typical platelet aggregation curve was obtained with collagen in the control cuvette, however excellent inhibition of platelet aggregations were obtained in the cuvette with PVC's containing PGI₂-Na.(Fig.2).

Fig.2. Platelet aggregation curves for PRP exposed to PVC with PGI₂-immobilized hydrogels.

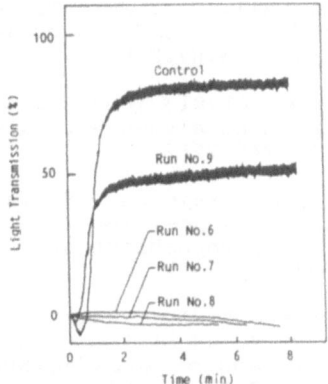

Next we followed the stability of this effect up to 46 h. The PVC surface containing acrylamide gel did not inhibit platelet aggregation at all, however 100 % of inhibitions of collagen-induced platelet aggregation were obtained for all the PVC's containing PGI₂-Na.-immobilized hydrogels for over 46 h.

The antithrombogenic effect of TM-immobilized glass beads was evaluated by the inhibiting effect of thrombin-induced platelet aggregation and prolongating effect of partial thromboplastin time(PTT). PTT with control glass was 112 second, and this was efficiently prolonged to 152 second with TM-immobilized glass. The effect of the TM-immobilized glass on platelet aggregation was evaluated through thrombin-induced platelet aggregation. Thrombin-induced platelet aggregation was inhibited in the presence of TM-immobilized glass. The inhibition of platelet aggregation was 18 % by 2.9 µg/ml of TM-immobilized glass in PRP and 42 % by 8.8 μg/ml respectively.

SUMMARY
Poly(vinyl chloride)(PVC) with heparin- and/or prostaglandin I_2 (PGI_2)-immobilized in crosslinked polyacrylamide hydrogels were prepared and evaluated their antithrombogenic properties through inhibiting activity of platelet aggregation and plasma clotting.
The heparin-immobilized PVC's exhibited excellent anticoagulant activity and the PGI_2 immobilized PVC's completely inhibited collagen-induced platelet aggregation for over 36 h.
Recombinant human thrombomodulin-immobilized glass beads also inhibited the thrombin-induced platelet aggregation and prolonged blood clotting time.

ACKNOWLEDGEMENT
The authors wish to thanks to Miss M. Uchikado for her technical supports. We also express our thanks to Prof. M. Osame for his continuous encouragement.

REFERENCES

1. Maruyama I.(1986) The regulation of blood coagulation by the endothelium. Acta Haematol Jpn. 49:146-153.

2. Baenziger N.L., Dillender M.J., and Majerus P.W.(1977) Cultured human skin fibroblasts and arterial cells produce a labile platelet-inhibitory prostaglandin. Biochem. Biophys. Res. Commun. 78:294-301.

3. Marcum J.A., Fritze L., Galli S.J.,Karp G.,and Rosenberg R.D. (1983) Microvascular heparin-like species with anticoagulant activity. Am. J. Physiol. 245:H725-H733.

4. Esmon C.T., and Owen W.G. (1981) Identification of an endothelial cell cofactor for thrombin-catalyzed activation of protein C. Proc. Natl. Acad. Sci. USA 78:2249-2252.

5. Maruyama I., Bell C.E., and Majerus P.W. (1985) Thrombomodulin is found on endothelium of arteries, veins, capillaries, and lymphatics, and on syncytiotrophoblast of human placenta. J. Cell Biol.101:363-371.

6. Akashi M., Takeda S., Miyazaki T., Yashima E., Miyachi T., Maruyama I., Okadome T., Murata Y. (1989) Antithrombogenic poly(vinyl Chloride) with heparin- and/or prostaglandin I_2-immobilized in hydrogels. J. Bioactive and Compatible Polym. 4:4-16.

7. Suzuki K., Kusumoto H., Deyashiki Y., Nishioka J., Maruyama I., Zushi M., Kawahara S., Honda G., Yamamoto S., and Horiguchi S.(1987) Structure and expression of human thrombomodulin, a thrombin receptor on endothelium acting as a cofactor for protein C activation. EMBO J. 6:1981-1987

8. Grode G.A., Pitman J., Crowley J.P.,Leininger R.I.and Falb R.D.(1974) Surface-immobilized prostaglandins as a platelet protective agent. Trans. Am. Soc. Artif. Intern. Organs. 20:38-41

9. Ebert C.D., Lee E.S., and Kim S.W.(1982)The antiplatelet activity of immobilized prostacyclin. J. Biomed. Mater. Res. 16:629-638

10.Jacobs H.A., Okano T., and Kim, S.W.(1989) Antithrombogenic surfaces:Characterization and bioactivity of surface immobilized PGE_1-heparin conjugate. J.Biomed. Mater. Res. <u>23</u>:611-630

11.Chandy T., and Sharma, C.P.(1984) The antithrombogenic effect of prostaglandin E immobilized on albuminated polymer matrix. J. Biomed. Mat. Res.<u>18</u>:1115-1124.

12.O'Brien, J.R.(1962) Platelet aggregation I. Some effects of the adenosine phosphates, thrombin, and cocaine upon platelet adhesiveness. J. Clin. Pathol.<u>15</u>:446-452

13.Gomi K., Zushi M., Honda G., Kawahara S.,Matsuzaki O., Kanabayashi T., Yamamoto S., Maruyama I.,and Suzuki K. (1990) Antithrombogenic effect of recombinant human thrombomodulin on thrombin-induced thromboembolism in mice. Blood <u>75</u>:1396-1399

HETEROGENOUS RESPONSES TO VASOACTIVE SUBSTANCES OF CANINE SUPERFICIAL AND JUXTAMEDULLARY AFFERENT ARTERIOLES

RYUICHI FURUYA, KAZUHISA OHISHI, AKIHIKO KATOH, AKIRA HISHIDA, NISHIO HONDA

The First Department of Medicine, Hamamatsu University School of Medicine, Hamamatsu, 431-31, Japan

INTRODUCTION

It is well known that there are structural and functional differences between superficial and juxtamedullary nephrons[1]. The size of glomeruli and the diameter of afferent arterioles are lager in juxtamedullary nephrons than in superficial nephrons. Also, the single nephron glomerular filtration rate (SNGFR) in the juxtamedullary nephron is usually 1.5 to 2.5 times greater than in the superficial nephron[2,3].

On the other hand, various procedures, such as saline loading[4], acute blood loss[5] and vasoactive substances[6,7] have been found to induce the intrarenal blood flow redistribution. As afferent arteriolar resistance plays an important role in altering glomerular blood flow, the intrarenal blood flow redistribution might be mediated through the heterogenous responses of afferent arterioles to vasoactive substances. In this regard, however, little imformation is available on the direct actions of vasoactive substances on afferent arterioles.

In this work, we examined direct actions of atrial natriuretic peptide(ANP), prostaglandin I_2(PGI$_2$), angiotensin II(AII), arginine vasopressin(AVP), meclofenamate(MF) and captopril(CAPT) on isolated perfused superficial and juxtamedullary afferent arterioles.

METHODS

Experiments were performed on 72 mongrel dogs. The afferent arteriole was dissected and perfused as described previously[8]. Briefly, a single afferent arteriole was dissected from either the most superficial 25% and the most juxtamedullary 25% of the canine renal cortex, and transferred to a temperature-controlled chamber and bathed in modified Ringer's solution bubbled with 95% O_2-5% CO_2. The afferent arteriole was drawn into holding pipette and held between holding pipette and perfusion pipette within the holding pipette. A pressure pipette filled with 5% Fast Green FCF was advanced into the arteriolar lumen through the perfusion pipette, to measure arteriolar pressure with Landis technique. A fourth pipette was used for the replacement and exchange of the perfusate. The distal end of the afferent arteriole remained open, and the outflow fluid drained into the bath (Fig.1).

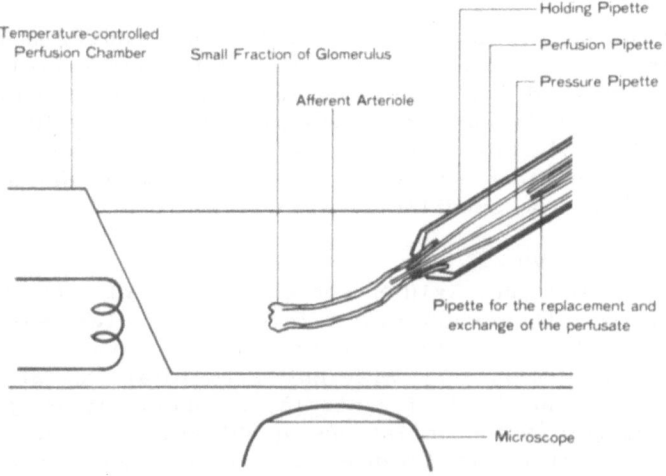

Fig.1. Perfusion system used for single afferent arteriolar perfusion

The bath medium and outflow fluid were collected at the end of control and experimental periods. Collected fluid volume was measured gravimetrically. The protein concentration in the perfusate and collected fluid was determined using modified Lowry method. The perfusion rate was calculated from protein concentration of the perfusate and collected fluid, collected fluid volume and the perfusion time. Arteriolar resistance was calculated from the perfusion rate and arteriolar pressure. Substances to be tested were added to the perfusate. Just before and after each experiment, the response to 1×10^{-6}M norepinephrine (NE) was examined to test the viability of the arteriole. When NE-induced increase in the arteriolar resistance was less than 500% of the initial value, the preparation was discarded and not studied.

For statistical analysis, one-way analysis of variance and the paired Student's t test were used. All data were expressed as mean ± SEM

RESULTS

Effects of ANP and PGI$_2$ on NE-contracted arterioles

Actions of ANP and PGI$_2$ were evaluated on NE-contracted afferent arterioles.

The 1×10^{-10}, 1×10^{-8} and 1×10^{-6}M concentrations of ANP significantly reduced the resistances of superficial and juxtamedullary arterioles(Fig.2). The percent reduction in the vascular resistance was significantly larger in juxtamedullary afferent arterioles than in superficial arterioles (p<0.01).

The 1×10^{-6}M concentration of PGI_2 did not significantly affect the resistance in both superficial and juxtamedullary arterioles, but PGI_2 at 1×10^{-6}M reduced the resistance in NE-contracted superficial and juxtamedullary afferent arterioles by approximately 40 and 64%, respectively(Fig.2). This percent reduction in the vascular resistance was significantly greater in the juxtamedullary afferent arterioles than in the superficial arterioles (p<0.05).

Effects of AII and AVP

The 1×10^{-12}, 1×10^{-9} and 1×10^{-6}M concentrations of AII did not significantly alter the superficial afferent arteriolar resistance. On the other hand, the 1×10^{-9} and 1×10^{-6}M concentrations of AII increased the juxtamedullary afferent arteriolar resistance to approximately 222 and 170 % of controls, respectively (Fig.3).

The 1×10^{-8} and 1×10^{-6}M concentrations of AVP did not significantly change the vascular resistance in both superficial and juxtamedullary afferent arterioles (Fig.3).

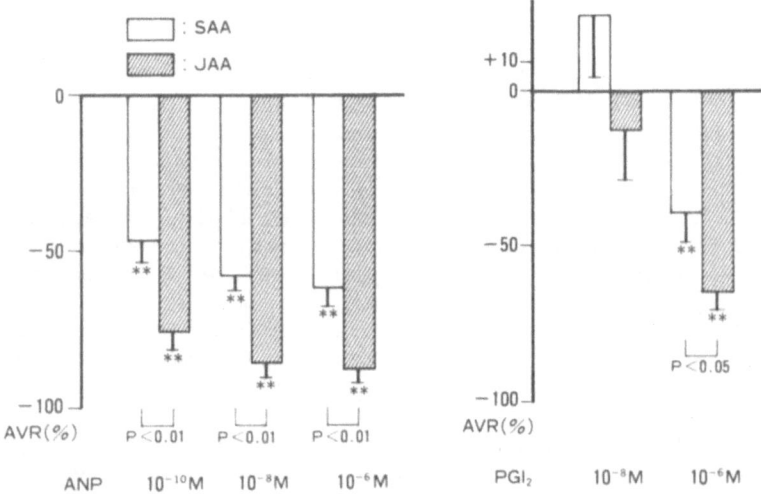

Fig.2. Effects of atrial natriuretc peptide (ANP) and prostaglandin $I_2(PGI_2)$ on NE(1×10^{-6}M)-contracted arterioles.
Data are expressed as percent change of arteriolar vascular resistance (AVR). NE, norepinephrine; SAA, superficial afferent arterioles; JAA, juxtamedullary afferent arterioles. N = 7 (ANP), 6 (PGI_2) in SAA and 5 (ANP), 5 (PGI_2) in JAA. ** p<0.01 compared with NE.

Effects of MF and CAPT

MF(1×10^{-5}M), a cyclo-oxygenase inhibitor, did not significantly affect the superficial arteriolar resistance, while provoked an

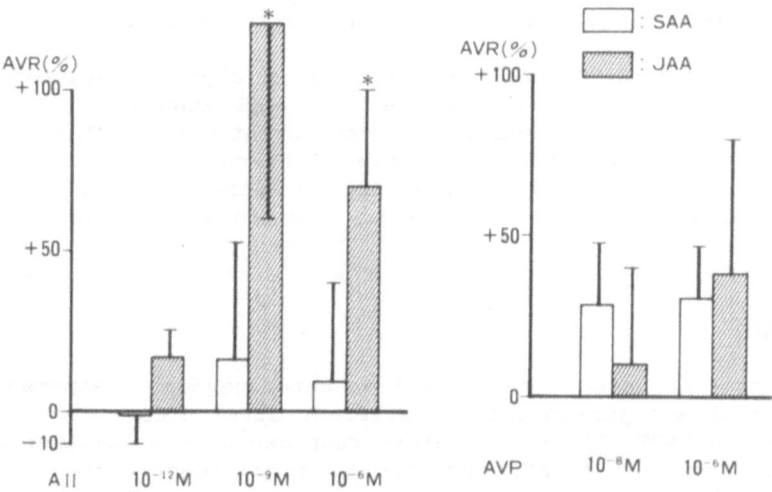

Fig.3. Effects of angiotensin II(AII) and arginine vasopressin(AVP).
Data are expressed as percent change of arteriolar vascular resistance (AVR). SAA, superficial afferent arterioles; JAA, juxtamedullary afferent arterioles. N = 5 (AII), 6 (AVP) in SAA and 7 (AII), 4 (AVP) in JAA. * p<0.05 compared with controls.

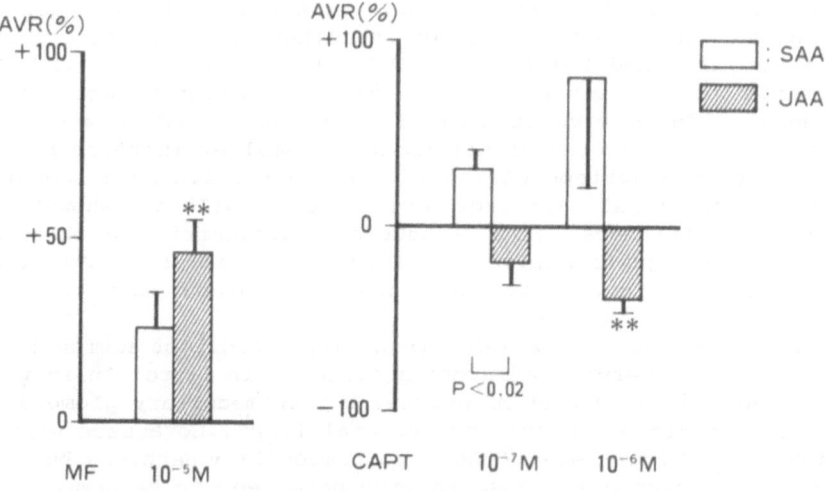

Fig.4. Effects of meclofenamate(MF) on NE-untreated and captopril(CAPT) on NE-contracted arterioles.
Data are expressed as percent change of arteriolar vascular resistance (AVR). SAA, superficial afferent arterioles; JAA, juxtamedullary afferent arterioles. N = 9 (MF), 5 (CAPT) in SAA and 7 (MF), 6 (CAPT) in JAA. ** p<0.01 compard with controls.

increase in the juxtamedullary arteriolar resistance by 46.3% (Fig.4).

The effects of CAPT (1×10^{-7} and 1×10^{-6}M), an angiotensin converting enzyme inhibitor, were examined on NE-contracted afferent arterioles. The 1×10^{-7} and 1×10^{-6}M concentrations of CAPT did not significantly affect the resistance of NE-contracted superficial arterioles. In contrast, the 1×10^{-6}M concentration of CAPT attenuated NE-treated juxtamedullary arteriolar resistance by 40% (Fig.4).

DISCUSSION

The present study demonstrated heterogeneous responses of superficial and juxtamedullary afferent arterioles to ANP, PGI_2, AII, MF and CAPT. The vasodilatory response of the juxtamedullary afferent arteriole to ANP and PGI_2 was significantly greater than that of the superficial arteriole. MF and CAPT reduced the resistance of the juxtamedullary afferent arteriole, but not in the superficial afferent arteriole. The data suggest that juxtamedullary afferent arterioles are more responsive to the vasoactive substances than superficial afferent arterioles.

Our data seem to be compatible with those by Edwards et al[9,10,11], Carmines et al[12] and Veldkamp et al[13]. Using isolated superficial afferent arterioles of rabbit, Edwards have demonstrated that PGI_2[9] has the direct vasodilatory action, while AII[10] and ANP[11] has no direct action on these vessels. On the other hand, Inscho et al[12], Carmines et al[13], and Veldkamp et al[14] noted that PGI_2[12], AII[13] and ANP[14] change significantly the juxtamedullary afferent arteriolar resistance. In in vivo studies, Steinhausen et al[15] showed that the intravenous injection of AII induces a smaller increase in the resistance of the juxtamedullary afferent arteriole when compared with the superficial afferent arteriole. Wilson[16] showed no significant difference in AII-induced vasoconstriction between superficial and juxtamedullary afferent arterioles. The true causes of the discrepancy between these in vivo and in vitro data are unclear.

Little imformation is available regarding the direct action of MF or CAPT on the glomerular afferent arterioles in vitro. An in vivo study has demonstrated that MF decreased juxtamedullary glomerular blood flow but did not change superficial flow[17]. Göranssen et al[18] showed that CAPT increased SNGFR in juxtamedullary nephrons but not in superficial nephrons. These in vivo data seem to be compatible with our findings in vitro.

In summary, there is a heterogeneity in responses to the vasoactive substances of superficial and juxtamedullary afferent arterioles in vitro. The responses to AII, MF and CAPT was observed only in juxtamedullary afferent arterioles. Furthermore, the responses to ANP and PGI_2 were greater in juxtamedullary afferent arterioles than in superficial arterioles. The intrarenal blood flow redistribution in various experimental conditions may be attributed, in part, to heterogeneous responses of afferent arterioles to vasoactive substances.

REFERENCES

1. Jacobson HR, Kokko PK (1985) The Kidney. Raven Press. New York.pp531-580
2. Horster M, Thurau K (1968) Pflügers Archiv 301: 162-181
3. Ericson AC, Sjöquist M, Ulfendahl HR (1982) Acta Physiol Scand 114: 203-209
4. Wallin JW, Blantz RC, Katz MA, Andreucci VE, Rector FC, Seldin DW (1971) Am J Physiol 221: 1279-1304
5. Rector JM, Stein JH, Bay WH, Osgood RW, Ferris TF (1972) Am J Physiol 222: 1125-1131
6. Bailie MD, Barbour JA (1975) Am J Physiol 228: 850-853
7. Britton SL (1981) Am J Physiol 240 (Heart Circ Physiol 9): H914-H919
8. Ohishi K, Hishida A, Honda N (1988) Am J Physiol 255 (Renal Fluid Electrolyte Physiol 24): F415-F420
9. Edwards RM (1985) Am J Physiol 248 (Renal Fluid Electrolyte Physiol 17): F779-F784
10. Edwards RM (1983) Am J Physiol 244 (Renal Fluid Electrolyte Physiol 13): F526-F534
11. Edwards RM, Weidley EF (1987) Am J Physiol 252 (Renal Fluid Electrolyte Physiol 21): F317-F321
12. Inscho EW, Carmines PK, Navar LG (1990) Am J Physiol 259 (Renal Fluid Electrolyte Physiol 28): F157-F163
13. Carmines PK, Morrison TK, Navar LG (1986) Am J Physiol 251 (Renal Fluid Electrolyte Physiol 20): F610-F618
14. Veldkamp PJ, Carmines PK, Inscho EW, Navar LG (1988) Am J Physiol 254 (Renal Fluid Electrolyte Physiol 23):F440-F444
15. Steinhausen M, Ballantyne D, Fretschner M, Hoffend J, Parekh N (1990) Kidney Int 38 (suppl.30):S55-S59
16. Wilson SK (1986) Kidney Int 30: 895-905
17. Kirschenbaum MA, White N, Stein JH (1974) Am J Physiol 227:801-805
18. Göransson A, Sjöquist M (1985) Acta Physiol Scand 124: 515-523

NATRIURETIC PEPTIDE mRNA IN SHR AND WKY

I.TANAKA[1], K.KOMATSU[1], T.FUNAI[2], A.ICHIYAMA[2] AND T.YOSHIMI[1]
[1]Second Department of Internal Medicine and [2]First Department of Biochemistry, Hamamatsu University School of Medicine, Hamamatsu, 431-31, Japan.

INTRODUCTION

Atrial natriuretic peptide (ANP) is a recently discovered hormone from the cardiac atria [1]. The potent diuretic, natriuretic and vasorelaxant activities of ANP suggested the involvement of this cardiac hormone in the regulation of blood pressure, body fluid and electrolytes. Originally, studies using radioimmunoassay (RIA) and blot hybridization technique revealed the presence of large quantities of immunoreactive ANP (IR-ANP) and ANP messenger RNA (ANP mRNA) in the atrium but not in the ventricle [2,3]. However, following studies indicated the synthesis of ANP in various extra-atrial tissues including the ventricle [4]. Its biological actions directed much attention to the implication of ANP in hypertension. Changes of IR-ANP concentrations in plasma and tissue of hypertensive rats were investigated [5,6]. Elevated ANP mRNA levels in atria and ventricles of spontaneously hypertensive rats (SHR) were also reported [7]. Recently, Sudoh et al. isolated and determined brain natriuretic peptide (BNP) from porcine brain, which derived from a distinct gene from ANP and possesses diuretic, natriuretic and vasorelaxant activities [8]. This natriuretic peptide family, ANP and BNP, seems to construct a complex regulatory system in mammalian circulation. In this study, to elucidate the pathophysiological role of natriuretic peptide family in hypertension, we tried to determine IR-ANP, IR-BNP and ANP mRNA levels in SHR and Wistar Kyoto rats (WKY).

MATERIALS AND METHODS

Animals
Male 17-week-old SHR and WKY (Charles River Co., Japan) were used. They were maintained in a temperature-humidity-light-controlled room with free access to regular rat chow MM-3 (Funabashi Co., Japan) and tap water ad libitum for 1 week before the experiments. Systolic blood pressure (BP) was measured using a tail-cuff manometer.

Tissue preparation
The tissues were removed immediately after decapitation. Hearts were dissected into ventricles and bilateral atria. To avoid contamination of the atrial tissue, the apical half of the ventricle was used for measurements of

IR-ANP, IR-BNP and ANP mRNA in the ventricle. The tissue was promptly
weighed, frozen and stored at -70 °C until use. Trunk blood was collected
into a plastic syringe containing aprotinin (500 kallikrein inactivator
units/ml) and Na$_2$-EDTA (1 mg/ml). Plasma was separated by centrifugation at 4
°C and immediately processed for extraction of ANP and BNP and subjected to
RIA as previously reported [9]. The tissues were boiled for 10 min in 10
volumes of 1 M acetic acid, then homogenized and centrifuged at 10,000 x g
for 30 min at 4 °C. The supernatants were lyophilized, reconstituted in the
assay buffer as previously described and subjected to RIA or gel filtration.
Gel filtration analysis was performed in a Sephadex G-75 column (1 x 56 cm)
equilibrated and eluted with the RIA buffer at the flow rate of 8 ml/hr.

RIA
Rat BNP-32 (Peptide Institution Inc., Japan) conjugated with bovine
thyroglobulin (Sigma, U.S.A.) using carbodiimide was emulsified with Freund's
complete or incomplete adjuvant and used for immunizing New Zealand White
rabbits. Rat BNP was radioiodinated by the chloramine T method and the
specific activity was 100 μCi/μg. Free and bound tracer were separated by the
polyethyleneglycol method. RIA for ANP was performed as previously described
[9].

RNA extraction and Northern blot analysis
RNA extraction was performed by Acid guanidinium-phenol-chloroform method.
RNA was determined spectrophotometrically with the assumption that absorbance
reading of 1.0 at 260 nm corresponds to a concentration of 40 μg/ml. ANP mRNA
level was determined by Northern blot analysis. An anti-sense RNA probe was
prepared by constructing a plasmid containing a Bgl II fragment,
corresponding to nucleotide residues 172 to 385, derived from rat ANP cDNA
[3] in the pGEM 3 (Takara, Japan). The plasmid DNA was digested by Eco RI
and used as a template for synthesis of ^{32}P-labeled RNA probe by SP6 RNA
polymerase. β-Actin mRNA was determined as an internal standard using β-actin
cDNA (Wako Junyaku Co. Ltd., Japan) labeled by random primer method with ^{32}P-
dCTP to specific activity of 1 x 10^9 cpm/μg. The relative ANP mRNA
concentrations were expressed in relation to the β-actin mRNA level and
values were normalized to the level of WKY.
All values were expressed as mean ± SEM. Student's t test was used for the
comparison of statistical significance.

RESULTS
The minimal detectable quantity of rat BNP was 1 fmol/tube and the
antiserum (final dilution 5 x 10⁴) did not recognize rat ANP, human BNP and
human ANP when as much as 100 pmol/tube of peptide was added.

There were significant differences in BP and ventricular weight between SHR and WKY. However, there were no differences in body weight and atrial weight (Table 1). Plasma IR-ANP concentration of SHR tended to be high, although there was no significant difference between SHR and WKY. however, plasma IR-ANP was not detectable in each group (less than 7 fmol/ml). Atrial and ventricular IR-ANP concentrations of SHR were 1.6 and 1.4 times higher than those of WKY. Atrial and ventricular IR-BNP concentrations of SHR were 1.5 times higher than those of WKY, each. The ratio of IR-BNP to IR-ANP (IR-BNP/IR-ANP) in atria was 4 % in SHR and WKY. IR-BNP/IR-ANP in ventricles was 46 % in SHR and 41 % in WKY (Fig. 1). Gel filtration chromatography revealed two components of IR-BNP in atrial extract of SHR and WKY. The big form (13K) was supposed to be the precursor protein and the small form (5K) eluted at the position of synthetic rat BNP-45, the latter of which was predominant in SHR and WKY. Similar pattern was observed in ventricular extract of SHR and WKY. The small form was the predominant component in SHR and WKY (Fig. 2). Northern blot analysis revealed that the atrial ANP mRNA of SHR was 121 % of WKY and that the ventricular ANP mRNA of SHR was 137 % of WKY, although these changes were not significant. The ANP mRNA contents of atrium and ventricle of SHR were significantly increased when compared with WKY, because total RNA from atrium and ventricle of SHR was increased (Fig. 3 and 4).

Table 1.

IR-ANP ánd IR-BNP in SHR and WKY

	SHR	WKY
Body weight (g)	324 ± 5.0	332.9 ± 3.8
Systolic blood pressure (mmHg)	172.2 ± 2.0**	118.2 ± 2.8
Atrial weight (mg)	50.8 ± 2.2	55.6 ± 2.4
Ventricular weight (mg)	1232 ± 39.2**	1087 ± 27.6
Plasma ANP (fmol/ml)	55.8 ± 6.8	43.5 ± 3.6
Plasma BNP (fmol/ml)	<7.0	<7.0
Atrial ANP (nmol/g tissue)	62.0 ± 3.7**	37.3 ± 2.8
Atrial BNP (nmol/g tissue)	2.3 ± 0.1**	1.5 ± 0.2
Ventricular ANP (pmol/g tissue)	23.1 ± 1.4**	16.2 ± 1.4
Ventricular BNP (pmol/g tissue)	6.0 ± 0.6*	4.0 ± 0.3

* $P<0.05$ **$P<0.01$ N=7

DISCUSSION

In this study, we demonstrated the presence of IR-BNP in rat atria and ventricles using a specific RIA for rat BNP. Gel chromatographic study revealed that IR-BNP in atria and ventricles was composed of two components

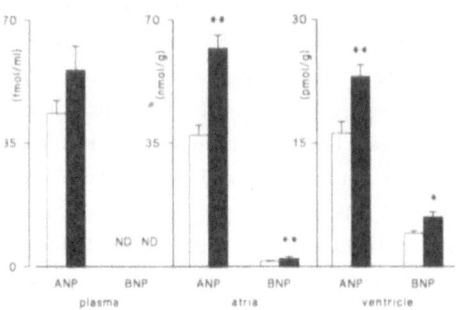

Fig. 1. Concentrations of IR-ANP and
IR-BNP in WKY (open column) and SHR
(black column).

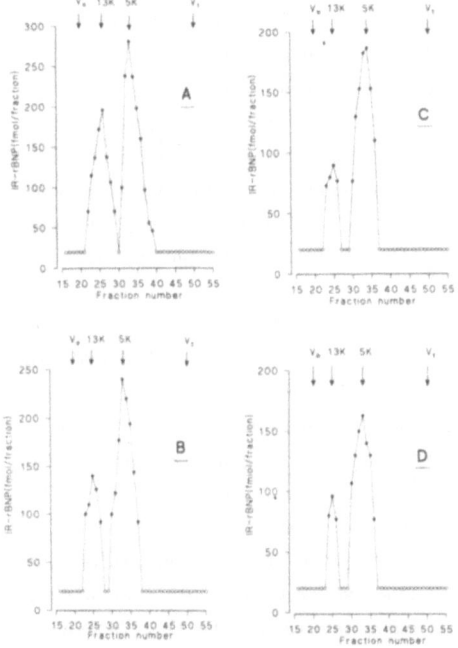

Fig. 2. Gel chromatographic profiles
of IR-BNP in atria of WKY (A), atria
of SHR (B), ventricles of WKY (C) and
ventricles of SHR (D).

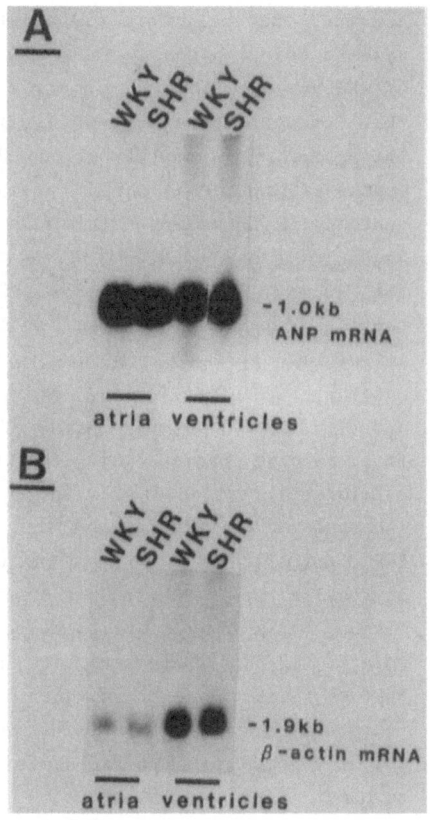

Fig. 3. Northern blot analysis.
A:ANP mRNA, B:β-actin mRNA.

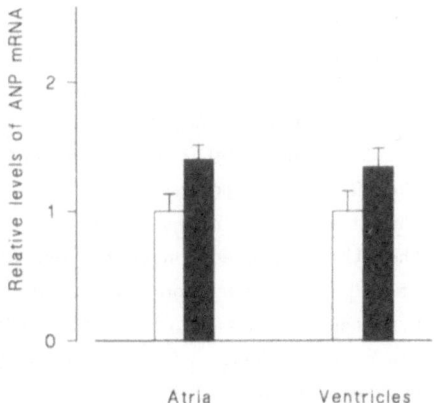

Fig. 4. ANP mRNA levels in WKY (open
column) and SHR (black column).

and that the small form eluted at the elution position of synthetic BNP-32 was the major component in atria and ventricles. In SHR, the concentrations of IR-BNP were increased in atria and ventricles when compared with those in WKY. However, there was not a difference in Ventricles/Atria of IR-BNP nor the processing of IR-BNP between SHR and WKY. We also confirmed the increased concentrations of IR-ANP in atria and ventricles of SHR and the increased contents of ANP mRNA in atria and ventricles of SHR.

Recent studies revealed the existence of 3 distinct natriuretic peptides, that is ANP, BNP and C type natriuretic peptide (CNP) [10], and natriuretic peptide receptor subtypes, A-, B- and C-receptor [11]. The relative affinities of these peptides to the 3 receptors have been studied. In rat natriuretic peptide family, BNP is absent in the central nervous system (CNS) and CNP is present only in the CNS, although BNP was originally discovered from porcine brain [12]. Therefore, ANP and BNP are the members of natriuretic peptide family in the peripheral tissue. Our study showed the increase of atrial and ventricular concentrations of IR-ANP and IR-BNP in SHR. However, the concentrations of IR-ANP were higher than those of IR-BNP suggesting that ANP is the major natriuretic peptide in the peripheral tissue. We were not able to detect plasma IR-BNP and the pathophysiological role of BNP in hypertensive rats was not clear in this study. Interestingly, BNP was relatively rich in ventricles and the presence of natriuretic peptide receptors in cardiac tissue has also been reported. These data may suggest the involvement of BNP in the development of cardiac hypertrophy as a local hormone.

Another interest is the pathopysiological roles of ANP in hypertension through the CNS. Existence of ANP specific receptors, IR-ANP and ANP mRNA in the rat brain has been demonstrated [2,13,14]. However, no consensus seem to have emerged yet concerning ANP level in the CNS of hypertension. We have examined IR-ANP and ANP mRNA level in the CNS of SHR using a RIA for ANP and the Ribonuclease protection assay (J. Hypertension in press). The concentrations of ANP mRNA in the hypothalamus and brainstem of SHR were 2.6 and 3.3 times higher than those of WKY (each P<0.05). The concentrations of IR-ANP in the hypothalamus and brainstem of SHR were 2.0 and 1.8 times higher than those of WKY (each P<0.05). Elevated mRNA levels in the CNS indicated that the increased concentrations of IR-ANP in the hypothalamus and brainstem of SHR resulted from increased synthesis of ANP. Considering the anti-hypertensive and antidipsogenic effects of centrally administered ANP, the elevated synthesis of ANP in the CNS seems to compensate the changes induced by hypertension.

Further studies using BNP cDNA obtained from rat atria by polymerase chain reaction are ongoing in our laboratory.

REFERENCES

1. Maki M. Takayanagi R. Misono KS. Pandey KN. Tibbetts C. Inagami T (1984) Nature 309:722-724

2. Tanaka I. Misono KS. Inagami T (1984) Biochem Biophys Res Commun 124:663-668

3. Nakayama K. Ohkubo T. Hirose S. Inayama S. Nakanishi S (1984) Nature 310:699-701

4. Takayanagi R. Imada T. Inagami T (1987) Biochem Biophys Res Commun 142:483-488

5. Tanaka I. Inagami T (1986) J Hypertens 4:109-112

6. Morii N. Nakao K. Kihara M. et al. (1986) J Hypertens 4(suppl 3):S317-S319

7. Arai H. Nakao K. Saito Y. et al. (1987) Circ Res 62:926-930

8. Sudoh T. Kangawa K. Minamino N. Matsuo H (1988) Nature 332:78-81

9. Morita H. Tanaka I. Oda T. Ichiyama A. Yamazaki T. Uematsu T. Nakashima M. Yoshimi T (1990) Peptides 11:843-847

10. Sudoh T. Minamino N. Kangawa K. Matsuo H (1990) Biochem Biophys Res Commun 168:863-870

11. Chinkers M. Gerbers DL. Chang MS. et al. (1989) Nature 338:78-83

12. Aburaya M. Minamino N. Hino J. Kangawa K. Matsuo H (1989) Biochem Biophys Res Commun 165:880-887

13. Brown J. Czarnecki A (1990) Am J Physiol 258:R1078-R1083

14. Gardner DG. Vlasuk GP. Baxter JD. Fiddes JC. Lewicki JA (1987) Proc Natl Acad Sci USA 84:2175-2179

DETERMINATION OF RENAL ALLOGRAFTS BLOOD FLOW BY ULTRASONIC DOPPLER METHOD

T.USHIYAMA

Department of Urology, Hamamatsu University School of Medicine, Hamamatsu, Japan.

Key words: ultrasonic Doppler method, renal transplantation, acute tubular necrosis, acute rejection, cyclosporine nephrotoxicity

INTRODUCTION

It is extremely important to establish an accurate noninvasive method of diagnosis of postoperative renal graft complications. Many diagnostic procedures have been used for this purpose. The author has studied the renal graft blood flow by the ultrasonic Doppler method to estimate the graft function.

MATERIALS AND METHODS

At the Hospital of Hamamatsu University School of Medicine, the Doppler flow technique was applied in 60 cases from February 1984 to June 1989. Twenty seven patients received grafts from living donors (LD) and 33 from cadaveric donors (CD). The study was performed using a directional Doppler flowmeter, Meda Sonics VASCURAB bidirectional Doppler Model D-10 (5MHz). Advance Angioscan was used for recording blood flow. In most cases, the examinations were carried out within one week, and then every one or two weeks after transplantation.

First, the ultrasonography was carried out to detect the renal graft artery. Then the probe was placed over the renal graft artery and the search was made for the strongest signal from the vessel. The renal graft blood flow was analyzed by the ratio of A/L to s ((A/L)/s; where A is the area of the waveform, L is the length (time) of the waveform, and s is the height of the waveform of the systolic phase)(Fig.1).

RESULTS

Among all patients, 37 had acute tubular necrosis (ATN, which necessitated hemodialysis), 19 acute rejection, and 8 cyclosporine (CyA) nephrotoxicity.

Good renal graft function

The author evaluated 26 examinations in 11 LD recipients without postoperative complications. The serum creatinine level (SCr) was a mean value of 1.5 mg/dl and the (A/L)/s ratio was a mean of 51%. There is no statistically significant correlation between the two groups (Fig.2). Among these cases, the lowest value of the (A/L)/s ratio was 42%. The author considered the ratio in well functioning grafts to be 45% or more.

Acute tubular necrosis (ATN)

In 37 ATN patients, 16 patients had no other complications. In 13 of 16 cases, the (A/L)/s ratio within one postoperative week was compared with ratio on the day of recovery from ATN. There was a statistically significant difference between the mean values in the two groups; 31%, 44%, respectively (p<0.002). In 10 cases, the ratios showed improvement as the renal function recovered. In 2 cases, the ratios showed

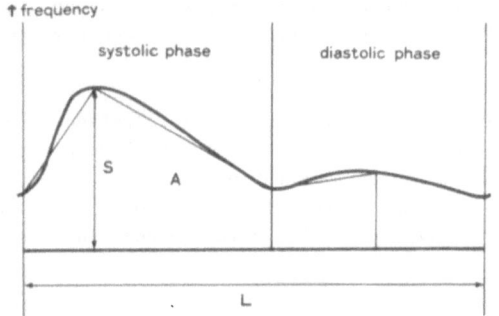

Fig.1. The waveform of the renal graft blood flow by the Doppler flow method. A, area of the waveform; L, length (time) of the waveform; s, height of the waveform of the systolic phase.

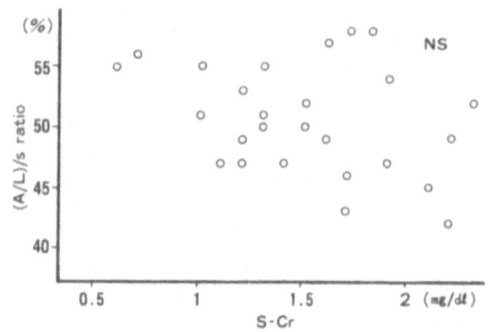

Fig.2. Relationship between serum creatinine level (SCr) and the (A/L)/s ratio. There is no statistically significant correlation between the two groups.

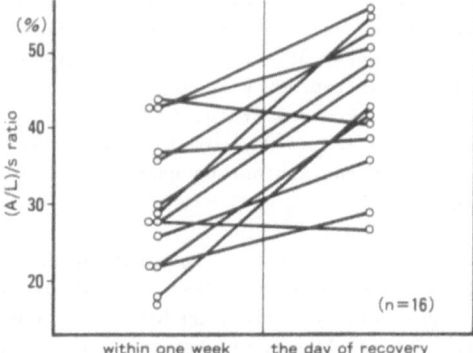

Fig.3. The change of the (A/L)/s ratio in ATN. In 13 of the 16 cases, the ratio within one week was compared with the ratio on the day of recovery from ATN.

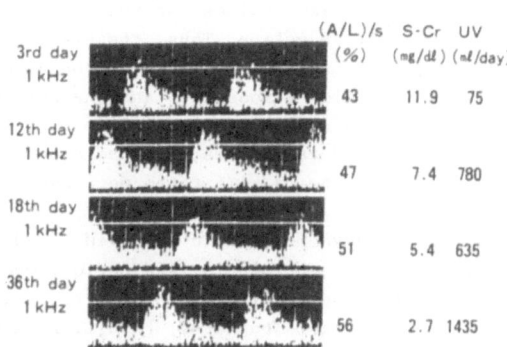

Fig.4. Sequential waveforms and the (A/L)/s ratios in the ATN patient (case 1).

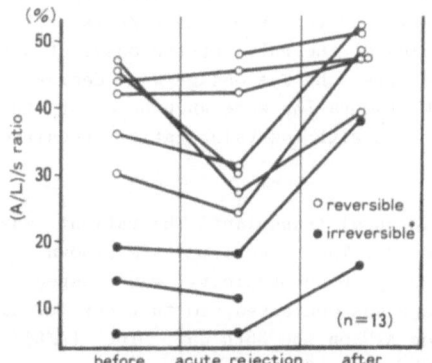

Fig.5. The change of the (A/L)/s ratio in acute rejection with ATN. *In 3 of the 6 irreversible cases, the renal graft blood flow could not be detected.

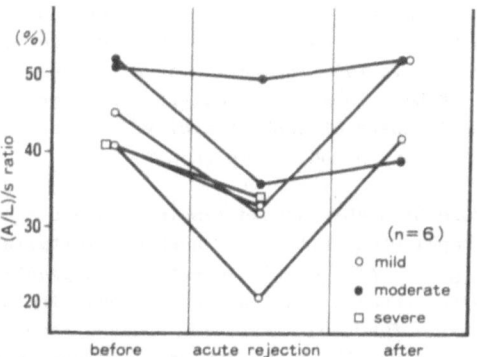

Fig.6. The change of the (A/L)/s ratio in acute rejection after recovery from ATN. Note that the ratios decreased in 5 of the 6 cases.

relatively high values within one week and were unchanged in ATN period. In one case, the ratio continued a low value in ATN period and improved after recovery from ATN (Fig.3).

Case 1: A 26-year-old male received a cadaveric renal transplant. This patient had no other complications in ATN period, and had recovered from ATN on the 33rd postoperative day. The waveforms are shown in Fig.4. The (A/L)/s ratio on the 3rd postoperative day was a good value of 43%. The ratio then gradually increased to 56% on the 36th postoperative day. As the ratio was improving, the urine volume increased and the serum creatinine level decreased.

Acute rejection
In 19 episodes of acute rejection, 13 were detected in ATN period, and 6 after recovery from ATN. Thirteen acute rejection episodes with ATN were classified into two groups: 7 reversible and 6 irreversible rejection episodes. In 4 of the 7 reversible episodes, the (A/L)/s ratios decreased at the onset of rejection, while those of remaining episodes were unchanged. In the 6 irreversible cases, the (A/L)/s ratios were low values, or renal graft blood flow could not be detected before diagnosis of rejection. Thus, these were suspected cases of acute vascular rejection (Fig.5). Six cases of acute rejection after recovery from ATN were graded into three groups: 3 mild, 2 moderate, and one severe case by recovery status from rejection. In all 3 mild cases recovering from ATN, the (A/L)/s ratios decreased at rejection. Of the 2 moderate cases, one had a decreased ratio and another unchanged. In the severe case, the (A/L)/s ratio decreased (Fig.6).

Case 2: A 33-year-old male received a cadaveric renal transplant. This patient had a hyperacute rejection in ATN period. On the 7th postoperative day, he developed fever, and the renal graft biopsy revealed hyperacute rejection. On his 10th day, the renal graft ruptured, resulting in graftectomy. The waveforms are shown in Fig.7. The (A/L)/s ratio was 14% on the 1st postoperative day and maintained a low value.

Cyclosporine (CyA) nephrotoxicity
In 8 CyA nephrotoxicity cases, 3 were found in ATN period and 5 after recovery from ATN. In the 3 ATN cases, two cases had continuously low values of the (A/L)/s ratio after transplantation and one had a decreased value of the ratio at the onset of nephrotoxicity. The ratios of all cases increased after the CyA dosage was decreased (Fig.8). In all 5 cases after recovery from ATN, the ratios were unchanged at the early phase of nephrotoxicity. Subsequently, 2 cases with low-value ratios required temporary hemodialysis (Fig.9).

Case 3: A 46-year-old female received a cadaveric renal transplant. The patient was complicated with CyA nephrotoxicity in ATN period. The waveforms are shown in Fig.10. ATN continued after transplantation and CyA nephrotoxicity was suspected after renal biopsy on the 21st day. The CyA dosage was decreased, so that her renal graft function improved, and she recovered from ATN on the 35th day. The (A/L)/s ratio was 26% on the 8th day and increased to 47% on the 37th day.

Prediction of reversibility of renal graft function
This study revealed that the (A/L)/s ratio was useful in evaluation of the renal graft recovering from ATN. The author then studied the prediction of renal graft

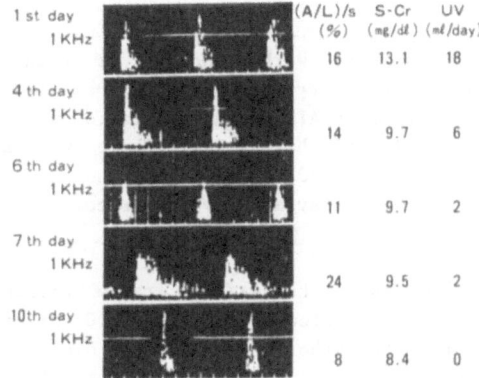

	(A/L)/s (%)	S-Cr (mg/dl)	UV (ml/day)
1st day 1KHz	16	13.1	18
4th day 1KHz	14	9.7	6
6th day 1KHz	11	9.7	2
7th day 1KHz	24	9.5	2
10th day 1KHz	8	8.4	0

Fig.7. Sequential waveforms and the (A/L)/s ratios in the patient with acute vascular rejection during ATN (case 2).

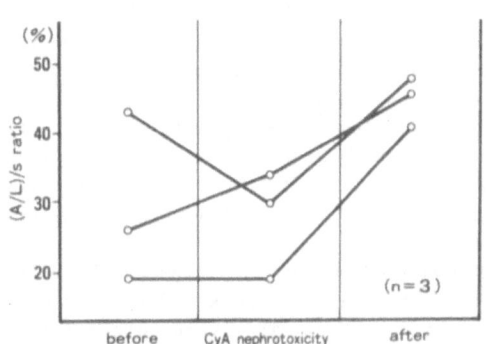

Fig.8. The change of the (A/L)/s ratio in CyA nephrotoxicity with ATN. All cases had a low ratio and recovered after the CyA dosage was decreased.

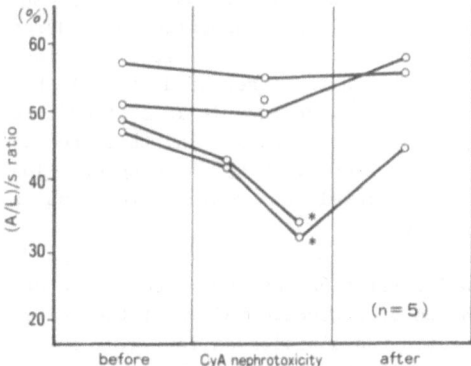

Fig.9. The change of the (A/L)/s ratio in CyA nephrotoxicity after recovery from ATN. *Two cases required hemodialysis.

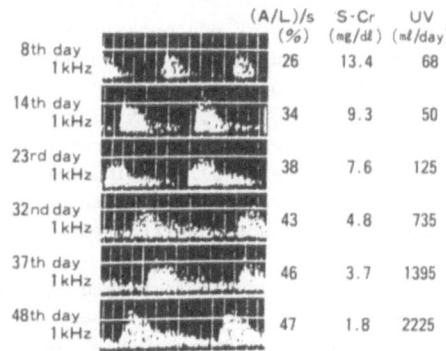

	(A/L)/s (%)	S-Cr (mg/dl)	UV (ml/day)
8th day 1kHz	26	13.4	68
14th day 1kHz	34	9.3	50
23rd day 1kHz	38	7.6	125
32nd day 1kHz	43	4.8	735
37th day 1kHz	46	3.7	1395
48th day 1kHz	47	1.8	2225

Fig.10. Sequential waveforms and the (A/L)/s ratios in the patient with CyA nephrotoxicity during ATN (case 3).

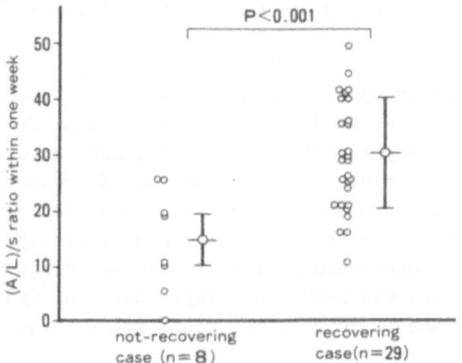

Fig.11. The prediction of the possibility of recovering from ATN. There is a statistically significant difference in the two groups (P<0.001).

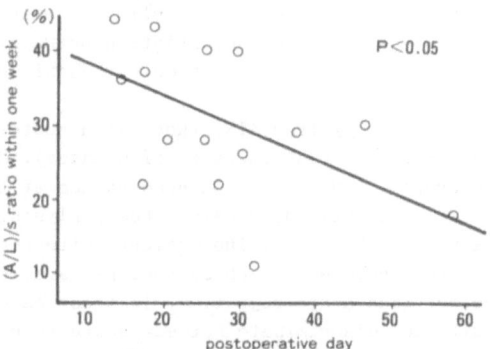

Fig.12. Relationship between the day of diuresis and the (A/L)/s ratio within one week. We assumed that diuresis was more than 1000ml of urine volume per day. The two groups had a statistical correlation (p<0.05).

function by the Doppler flow method.

First, the author studied whether the value of this ratio within one week of transplantation was useful in the prediction of the possibility of recovering from ATN. In 37 patients with ATN, 8 patients not recovering from ATN had a mean ratio value of 15%. The other 29 patients recovering from ATN had that of 30%. There was a statistically significant difference in the two groups (p<0.001)(Fig.11). The cases with the ratios of 20% or less on the 7th postoperative day may have other complications than ATN, especially acute rejection.

Secondly, the author studied the relationship between the (A/L)/s ratio within postoperative one week and the day when patients presented diuresis in 16 cases recovering from ATN without other complications. The author assumed that diuresis was more than 1000ml of urine volume. The two groups had a statistical correlation (p<0.05)(Fig.12). If the patients had a longer period of ATN than days predicted by the ratio, they may suffer other complications.

DISCUSSION
In renal transplantation, it is important that early diagnosis is made and prompt therapy is given for postoperative complications. Multiple diagnostic procedures have been performed, but no single method except renal biopsy is available for diagnosis of rejection, especially in the period of ATN. On the other hand, as CyA has been widely used, CyA nephrotoxicity has been recognized. Besides it is difficult to distinguish CyA nephrotoxicity from acute rejection in some cases. In this retrospective study, the (A/L)/s ratio of the renal graft blood flow by the Doppler flow method did not correlate well with serum creatinine level, but consecutive studies were useful to diagnose posttransplant complications.

In ATN, the (A/L)/s ratio increased as the renal graft function recovered. In acute rejection with ATN, reversible cases had unchanged or decreased values of the ratio. On the other hand, in irreversible cases, the ratio was a prolonged low value or no blood flow was detected. In CyA nephrotoxicity complicated with ATN, the ratio was a continuously low value after transplantation, or decreased at the onset of nephrotoxicity. In the early postoperative period, the ATN patients with prolonged low values of the ratio were complicated most likely with acute rejection or CyA nephrotoxicity, but the differentiation between acute rejection and CyA nephrotoxicity was difficult by the Doppler flow method alone.

After recovery from ATN, those with acute rejection episodes had decreased values of the ratio in most cases (5 of 6 cases). On the other hand, all 5 patients with CyA nephrotoxicity had unchanged values at the early phase of the toxicity, and 2 of these patients who received hemodialysis for the severe toxicity showed decreased values of the ratio. The patients with renal graft dysfunction and unchanged ratios in the recovering phase from ATN are more likely to suffer CyA nephrotoxicity, rather than acute rejection. In those cases with decreased ratios, it may be difficult to discriminate between acute rejection and CyA nephrotoxicity, but in CyA nephrotoxicity cases, the ratio may not decrease as long as the renal graft function is not severely deteriorated.

Patients with ATN have abnormal waveforms in some reports [1,2], and in others [3,4] they have normal waveforms. As to acute rejection, most reports reveal an increase in vascular resistance of renal graft [1,2,3,4]. In a few reports [5], patients with acute cellular rejection have normal waveforms. In CyA nephrotoxicity, the renal blood flows are reported to range from unchanged [3,5] to decreased [6]. Most reports have used the resistive index [2,3,4], or pulsatility index [1] to evaluate the waveform, and have estimated the renal graft function by the value of the index. These indices are considered to represent the peripheral resistance. We investigated more exactly the parameters that expressed the renal blood flow and used the area of the waveform ("A"), which was regarded as a parameter correlated with blood flow volume [7]. To neglect the heart rate and the angle between the Doppler probe and the vessel, "A" was divided by the length of the waveform ("L") and the height of the waveform of the systolic phase ("s"), respectively. Then using the (A/L)/s ratio, the author evaluated the waveform by the Doppler flow method.

The interstitial cell infiltration and edema, and vascular collapse may induce an abnormal blood flow in ATN. In acute vascular rejection, the renal blood flow is severely decreased by thrombus and vasculitis. In acute cellular rejection, as interstitial cell infiltration and edema progress, the renal blood flow is disturbed to various extents. In CyA nephrotoxicity, tubular degeneration is mainly found, so that the blood flow is not disturbed in the early phase. However, as the toxicity is developed, interstitial edema, fibrosis and vascular change lead to a decreased blood flow.

From this study, the author concludes that the postoperative monitoring of the renal graft blood flow by the Doppler flow method is useful for indicating recovery from ATN and for detecting complications in ATN period. However, acute rejection and CyA nephrotoxicity with ATN cannot be differentiated by this method alone. After recovery from ATN, we may be able to differentiate acute rejection from CyA nephrotoxicity by the Doppler flow method in most cases. Moreover, this procedure is useful in estimation of the possibility of recovering from ATN and prediction of the duration of the postoperative ATN. The author would like to emphasize that it is important to study the sequential change of the renal blood flow by the Doppler flow method.

Acknowledgments. The author is very grateful to Professor Kazuki Kawabe, Department of Urology, Hamamatsu University School of Medicine, Associate Professor Atsushi Tajima and Professor Yoshio Aso, Department of Urology, The University of Tokyo, the Faculty of Medicine for their guidance and encouragement during this study.

REFERENCES
1. Rigsby CM, Burns PN, Weltin GG, Chen B, Bia M, Taylor KJW (1987) Radiology 162: 39-42
2. Don S, Kopecky KK, Filo RS, Leapman SB, Thomalla JV, Jones JA, Klatte EC (1989) Radiology 171: 709-712
3. Rifkin MD, Needleman L, Pasto ME, Kurtz AB, Foy PM, McGlynn E, Canino C, Baltarowich OH, Pennell RG, Goldberg BB (1986) AJR 148: 756-762
4. Wan SKH, Ferguson CJ, Cochlin DLL, Evans C, Griffiths DFR (1989) Clinical Radiology 40: 573-576
5. Buckley AR, Cooperberg PL, Reeve CE, Magil AB (1987) AJR 149: 521-525
6. Ubhi CS, Guillou PJ, Irving HC, Giles GR, Norwood MK (1987) Annals of the Royal College of Surgeons of England 69: 229-232
7. Ueshima T (1974) Adult Diseases 14: 61-87

PLASMA LEVELS OF THROMBOMODULIN AND ACTIVE PLASMINOGEN ACTIVATOR INHIBITOR INCREASE IN DISSEMINATED INTRAVASCULAR COAGULATION WITH MULTIPLE· ORGAN FAILURE

H.ASAKURA, H.JOKAJI, M.SAITO, C.UOTANI, I.KUMABASHIRI, E.MORISHITA, M.YAMAZAKI AND T.MATSUDA

Department of Internal Medicine (III) , Kanazawa University School of Medicine, 13- 1 Takaramachi, Kanazawa, Japan

INTRODUCTION

Disseminated intravascular coagulation (DIC) is a serious clinical condition, but its pathophysiology is very different according to the underlying diseases responsible for DIC. For example, a marked activation of the coagulation and fibrinolytic system is present in most cases of acute promyelocytic leukemia. On the contrary, a little activation of the fibrinolytic system is present in most cases of sepsis with DIC[1-4]. Therefore, it is important to investigate these pathologic condithions separately according to the underlying disease responsible for DIC.

Thrombomodulin on the vascular endothelial cell surface plays an important role as a cofactor in catalyzed activation of protein C by thrombin-thrombomodulin complex, and activated protein C functions as an anticoagulant by inactivating the coagulaion foctor V a and VIIIa[5,6]. Thrombomodulin is also known to be present in plasma[7]. Since thrombomodulin is leberated from vascular endothelial cells into circulating blood by the damage to the cells, quantitative assay of soluble thrombomoulin in plasma seems to be useful to assess the degree of the damage to vascular endothelial cells[8].

Plasminogen activator inhibitor type 1 (PAI- 1) , which is also liberated from vascular endothelial cells, is a specific inhibitor of tissue-type plasminogen activator (t-PA)[9-10]. Active PAI, which has the capacity to bind to free t-PA, reflects the intensity of inhibitory regulation on fibrinolytic system in vivo[10].

In this paper, we examined the changes in plasma levels of soluble thrombomodulin and active PAI in cases of DIC to investigate the damage to vascular endothelial cells and the fibrinolytic balance. The relationship among multiple organ failure (MOF) , plasma levels of thrombomodulin and active PAI is also discussed in cases of DIC.

PATIENTS, MATERIALS AND METHODS

Patients The analysis of thrombomodulin and active PAI was performed in 66 patients with DIC. They were classified according to the underlying diseases responsible for DIC, as follows : 21 patients with acute promyelocytic leukemia, 20 with acute leukemia except acute promyelocytic leukemia, 4 with blastic crisis of chronic myelogenous leukemia, 8 with

TABLE 1 Diagnosis of MOF *

| |
If the paitients had two or more of the following disorders, the existence of multiple organ failure is highly possible.

Heart *
1. Acute myocardial infarction
2. Arrhythmia (bradycardia, heart rate <50/min ; ventricular tachycardia ; ventricular fibrillation ; atrioventricular block ; cardiac arrest)

Lung *
1. Pao <50mmHg
2. Requirement of artificial respiration

Kidney *
1. Serum creatinine≧3mg/dl (N:0.5—1.2mg/dl)
2. Serum BUN≧50mg/dl (N:3—21mg/dl)

Liver ᵇ
1. Serum total bilirubin≧3mg/dl (N:0.36—1.30mg/dl)
2. Serum GOT (N: 10—40 IU/L) and GPT (N: 3—47 IU/L) ≧100 IU/L
3. LDH≧800 IU/L (N: 205—415 IU/L)

Gastrointestinal tract *
1. Hematemesis and the origin of bleeding confirmed by endoscopy
2. Melena and the origin of bleeding confirmed by endoscopy

Brain ᶜ
1. Only response to painful stimuli

*Presence of one or more of the listed disorders. ᵇPresence of two or more of the listed disorders. ᶜPresence of the following. *N=normal values

non—Hodgkin lymphoma, 7 with solid cancer and 6 with sepsis.

Diagnosis of disseminated intravascular coagulation (DIC)   The diagnosis of DIC was made based on the criteria proposed by Research Committee on DIC in the Ministry of Health and Welfare of Japan [11].

Diagnosis of multiple organ failure (MOF)   The diagnosis of multiple organ failure in patients with DIC was made with the criteria as shown in Table 1.   Thirty of 66 cases were diagnosed as having multiple organ failure : 13 without renal failure and 17 with renal failure.

Assay methods   Prothrombin time (PT) was measured using an automatic recorder (Auto—Fi, Dade) .   Plasma fibrinogen was determined by Clauss' indirect method and serum concentration of fibrin/fibrinogen degradation products (FDP)   by the latex agglutination immunoassay.   Plasma levels of thrombin—antithrombin Ⅲ complex were measured using an ELISA kit (Behringwerke) and those of plasmin— α_2 plasmin inhibitor (or α_2 antiplamin)   complex, using an ELISA kit (Teijin Ltd.) [13].   Plasma levels of tissue plasminogen activator (t—PA) —plasmingen activator inhibitor (PAI—1) complex were measured by a two step sandwich ELISA using polyclonal antibody against human PAI—1 and peroxidase—conjugated polyclonal anti—t—PA antibody (Teijin Ltd.) [10].   Plasma levels of active PAI, which has the capacity to bind to free t—PA, were calculated from the increment of t—PA · PAI—1 complex following the addition of excess t—PA to plasma. Plasma levels of soluble thrombomodulin were measured by one step sandwich ELISA using polyclonal antibody against human thrombomodulin and peroxidase—conjugated monoclonal anti—thrombomodulin antibody, which recognizes around thrombin binding site.

Blood samples   Blood samples were obtained from the patients by antecubital venipuncture, when the diagnosis of DIC was made.   In some of 66 cases of DIC, the analysis was repeated 6 to 8 days after the diagnosis of DIC.   The blood samples were anticoagulated

TABLE 2 Laboratory findings in patients with DIC

Parameters	Normal values	APL (n=21)	ALsAPL (n=20)	CML · BC (n=4)	NHL (n=8)	Cancer (n=7)	Sepsis (n=6)
PT (sec)	11.9−13.9	17.1±3.9	14.4±2.1	14.5±2.1	13.9±2.1	14.1±3.2	16.2±4.5
Fbg (mg/dl)	170−410	104.4±51.6	234.4±98.9	178.0±96.2	240.3±25.7	284.8±90.4	244.4±168.1
FDP (μg/ml)	<5.0	75.9±54.3	60.1±81.8	53.0±38.2	59.1±41.7	102.3±67.7	39.0±26.2
TAT (ng/ml)	<3.0	31.5±19.7	16.2±12.7	10.6±0.8	17.4±13.7	21.6±14.3	29.3±26.7
PIC (μg/ml)	<0.8	9.3±6.0	4.3±2.0	1.8±1.4	2.0±1.0	6.5±4.2	1.9±1.0
tPA/PAI (ng/ml)	5.5−16.1	14.0±6.7	26.2±17.8	54.5±48.4	29.8±21.2	21.5±4.7	48.8±22.4
ActivePAI (ng/ml)	3.2−19.0	18.2±42.0	40.6±36.0	68.0±38.5	30.9±45.7	30.0±34.7	115.0±42.8
TM (ng/ml)	3.4−13.0	12.1±8.7	18.5±12.4	24.1±15.5	16.0±3.5	16.9±9.4	33.1±11.4

PT, prothrombin time ; Fbg, fibrinogen ; FDP, fibrin/fibrinogen degradation products ;
TAT, thrombin−antithrombin Ⅲ complex ; PIC, plasmin− α plasmin inhibitor complex ;
tPA/PAI, t−PA · PAI−1 complex ; TM, throbomodulin.
  APL, acute promyelocytic leukimia ; ALsAPL, acute leukimia except APL ;
CML · BC, blastic crisis of chronic myelogenous leukemia ; NHL, non−Hodgkin lymphoma

with 3.8% sodium citrate (9/ 1, vol/vol, blood/anticoagulant) and centrifuged at 2,000g
for 10 minuted at 4°C immediately after venipuncture. Plasma was separated and stored
at − 70°C until use.

Statistical analysis The significance of difference in the values obtained from the patients
with different underlying diseases responsible for DIC was analysed with the nonparametric
Wilcoxon signed rank test. Correlation coefficient was calculated by linear regression
analysis.

RESULTS

Table. 2 shows laboratory findings including thrombomodulin and active PAI in 66 patients
with DIC. A significant elevation of plasma levels of solule thrombomodulin and active
PAI was observed in most cases of DIC, especially those with sepsis or blastic crisis ofch
ronic myelogenous leukimia. However, there was no significant elevation in most cases
of acute promyelocytic leukemia.

  Plasma levels of soluble thrombomodulin were directly proportional to plasma levels of
active PAI in 66 cases of DIC ($r=0.612$, $p <0.001$).

  Plasma levels of soluble thrombomodulin and active PAI in the cases of DIC with or
without multiple organ failure were compared as shown in Table. 3. Plasma levels of
soluble thrombomodulin in patients with multiple organ failure were significantly higher than
in those without multiple organ failure ($p <0.01$). No significant difference in plasma
levels of soluble thrombomodulin was observed between the cases of multiple organ failure
with renal failure and those without renal failure. Plasma levels of active PAI in
patients with multiple organ failure were significantly higher than in those without multiple
organ failure ($p <0.01$). There was no significant difference in plasma levels of active

TABLE 3 Thrombomodulin, Active PAI and multiple organ failure

| | MOF (−) | MOF (+) | |
| | | RF (−) | RF (+) |
	(n=36)	(n=13)	(n=17)
TM (ng/ml)	10. 4±4. 8	23. 9±12. 3	28. 2±10. 7
Active PAI (ng/ml)	9. 9±9. 2	61. 9±38. 9	85. 7±54. 8

TM, thrombomodulin ; MOF, multiple organ failure ;
RF, renal failure.

PAI between the cases of multiple organ failure with renal failure and those without renal failure.

Plasma levels of soluble thrombomodulin and active PAI were decreased with the clinical improvement in most cases of DIC with some exceptions observed particularly in the cases who could not be restored from DIC and died (Fig.1) . One patient, who died in spite of normal plasma levels of thrombomodulin and active PAI, was a case of acute promyelocytic leukemia without maltiple organ failure in whom the cause of his death was cerebral bleeding.

DISCUSSION

Thrombomodulin exists on the vascular endothelial cell surface and plays an important role in an anticoagulant effect of endothelium. Soluble thrombomodulin is also present in plasma. Since it has been reported that soluble thrombomodulin in plasma is liberated from endothelial cells in response to damage to these cells[7,8], it would be reasonable to regard that increased levels of plasma thrombomodulin reflect endothelial cell damage. A significant elevation of plasma levels of soluble thrombomodulin was observed in most cases of DIC with some exceptions. This result suggests that the damage to endothelial cells does commonly extist in most cases of DIC. It is interesting that no such significant elevation of plasma levels of soluble thrombomodulin was observed in most cases of acute promyelocytic leukemia, because it is well known that most serious DIC is accompanied with acute promyelocytic leukemia. Therefore, excess thrombin or plasmin generated in circulating blood does not seem to be a so important stimulant to liberate thrombomodulin from endothelium.

PAI− 1 is also liberated from endothelial cells, and plays an important role in the inhibition of fibrinolytic system as a specific inhibitor of t−PA. Some PAI exists as a complex with t−PA (t−PA · PAI− 1 complex) , and the other PAI exists as a free PAI (latent PAI or active PAI) in plasma. Active PAI, which has the capacity to bind to free t−PA, reflects the intensity of inhibitory regulation on fibrinolytic system in vivo. A significant elevation of plasma levels of active PAI was observed in most cases of DIC with a significant elevation of plasma levels of t−PA · PAI− 1 complex. Since plasma levels

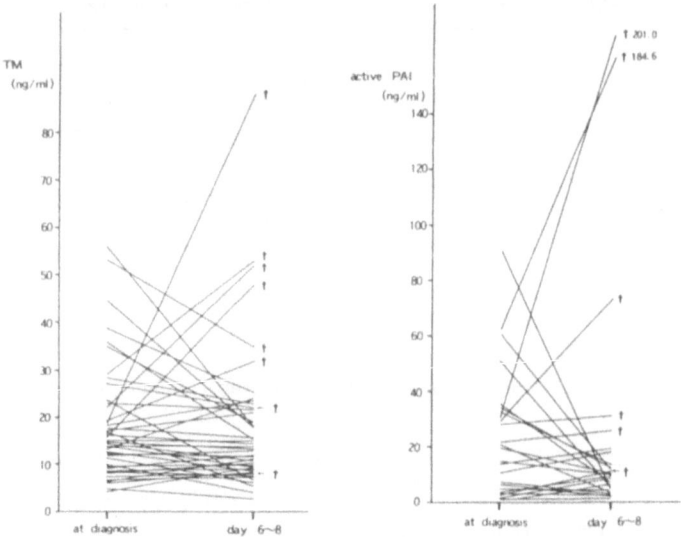

Fig. 1. Plasma levels of thrombomodulin and active PAI in cases of DIC at diagnosis and 6 to 8 days after diagnosis of DIC. ✝ , patients with DIC who could not be restored from DIC and died.

of active PAI were directly proportional to plasma levels of t−PA · PAI− 1 complex [4], the activation and inhibition of fibrinolytic system were well balanced in the cases of DIC. In cases of APL, no significant elevation of active PAI was observed, which may explain the extremely activated fibrinolysis in APL[1,2].

Multiple organ failure is a serious clinical condition and is well known as an important complication in DIC. As shown in Table. 3, plasma levels of soluble thrombomodulin were higher in the cases of DIC with multiple organ failure than in those without multiple organ failure, which suggests that the damage to endothelial cells was more prominent in the former than in the latter. It is possible that decrease in thrombomodulin on vascular endothelial cells by its liberation may have some influence on further progression of multiple organ failure in DIC. It has been reported that most of the plasma soluble thrombomodulin is catabolized in the kidney and that plasma levels of soluble thrombomodulin are also affected by renal dysfunction. However, no significant difference of plasma levels of soluble thrombomodulin was observed between the patients without renal failure and those with renal failure in the cases of DIC with multiple organ failure. Plasma levels of active PAI were higher in the cases of DIC with multiple organ failure than in those without multiple organ failure, which suggests that the inhibitiion of fibrinolytic system was more prominent in the former than in the latter. It is possible that the disturbance of microcirculation caused by the inhibition of fibrinolytic system may have some influence on further progression of multiple organ failure in DIC. Since both plasma levels of thrombomodulin and active PAI were significantly increased in the cases of DIC with multiple organ failure and a significantly positive correlation was observed between these parameters, the damage to endothelial cells due to circulatory failure would accelarate release of TM from these cells.

Plasma levels of thrombomodulin and active PAI were gradually decreased with the improvement of DIC, as shown in Fig. 1. On the other hand, in the patients who showed no improvement of DIC, both parameters remained at high levels or further increased. This suggests that both parameters may be useful to predict the prognosis of DIC.

In conclusion, both plasma levels of thrombomodulin and active PAI were increased in most cases of DIC, especially in those with sepsis or blastic crisis of chronic myelogenous leukemia, and this elevation was more evident in the cases of DIC with MOF. On the contrary, most cases of APL had no elevation in plasma levels of trombomodulin and active PAI. It would be useful to measure plasma levels of thrombomodulin and active PAI for evaluation of clinical conditions of DIC.

REFERENCES

1. Takahashi H., Hanano M., Takizawa S., Tatewaki S. and Shibata A. (1988) Plasmin−α_2−plasmin inhibitor complex in plasma of patients with disseminated intravascular coagulation. Am J Haematol 28:162−166.

2. Matsuda T., Asakura H., Ito K., Saito M., Jokaji H., Uotani C. and Kumabashiri I. (1988) Changes in levels of t−PA and α_2PI−plasmin complex in plasma in patients with DIC. Thromb Res [Suppl] 7:143−151.

3. Asakura H., Saito M., Ito K., Jokaji H., Uotani C., Kumabashiri I. and Matsuda T. (1988) Levels of thrombin−antithrombin III complex in plasma in cases of acute promyelocytic leukemia. Thromb Res 50:895−899.

4. Asakura H., Jokaji H., Saito M., Uotani C., Kumabashiri I., Morishita E., Yamazaki M. and Matsuda T. (1991) Changes in plasma levels of tissue−plasminogen activator/inhibitor complex and active plasmingen activator inhibitor in patients with disseminated intravascular coagulation . Am J Haematol 36:176−183

5. Esmon C.T. and Owen W.G. (1981) Identification of an entothelial cell cofactor for thrombin−catalyzed activation of protein C. Proc Natl Acad Sci USA 78:2249−2252

6. Salem H.H., Mruyama I., Ishii H. and Majerus P.W. (1984) Isolation and characterization of thrombomodulin from human placenta. J Biol Chem 259:12246−12251.

7. Ishii H. and Majerus P.W. (1985) Thrombomodulin is present in human plasma and urine. J Clin Invest 76:2178−2181.

8. Takano S., Kimura S., Ohdama S. and Aoki N. (1990) Plasma thrombomodulin in health and diseases. Blood 76: 2024– 2029.

9. Kruithof E.K.O., Tran T.C., Ransijn A. and Bachmann (1984) Demonstration of a fast–acting inhibitor of plasminogen activators in human plasma. Blood 64: 907–

10. Sakata Y., Murakami T., Noro A., Mori K. and Matsuda M. (1991) The specific activity of plasminogen activator inhibitor– 1 in disseminated intravascular coagulation with acute promyelocytic leukemia. Blood 77: 1949– 1957

11. Kabayashi N., Maekawa T., Takada M. et al (1983) Criteria for diagnosis of DIC based on the analysis of clinical and laboratory findings in 345 DIC patients collected by the Research Committee on DIC in Japan. Bibliotheca Haematologica 49: 265– 275

12. Pelzer H., Schwarz A. and Heimburger N. (1988) Determination of human thrombin– antithrombin Ⅲ complex in plasma with an enzyme–linked immunosorbent assay. Thromb Haemost 59: 101– 106.

13. Mimuro J., Koike Y., Sumi Y. and Aoki N. (1987) Monoclonal antibodies to discrete regions in α_2–plasmin inhibitor. Blood 69: 446– 453.

SIGNIFICANCE OF PLASMINOGEN ACTIVATORS AND PLASMINOGEN ACTIVATOR INHIBITORS IN HUMAN LUNG CANCER

M.NAGAYAMA, H.HAYAKAWA, A.SATO, T.URANO*, Y.TAKADA*, AND A.TAKADA*
Second Department of Internal Medicine and Second Department of Physiology*, Hamamatsu University, School of Medicine, Hamamatsu-shi, Shizuoka-ken, 431-31 Japan

INTRODUCTION

Plasminogen activators (PAs) are serine proteases which convert plasminogen to plasmin and play a major role to regulate intravascular fibrinolysis and extracellular proteolysis.[1] At least two types of plasminogen activator, tissue plasminogen activator (t-PA), urokinase (UK), have been identified. UK and its precursor, pro-UK, which are released from cancer cells and bind to specific cell surface receptor of these cells,[2-3] is considered to play important roles in tumor invasion or metastsis by regulating the degradation of the extracelluler matrix.

Plasminogen activator inhibitors (PAIs) are serine proteases inhibitors which specifically inhibit the activity of plasminogen activators. Three types of PAI, that are PAI-1, PAI-2, and PAI-3, are identified. Recently it has been reported that the activity and expression of PAI-1 and PAI-2 are important in tumorigenesis and the mechanism of metastasis.[4-5]

In the present study we measured the antigen levels of PAs and PAIs in lung cancer and tried to assess physiological and pathological relevance of these parameters.

MATERIALS AND METHODS

Tissue extract samples
Tumor specimens were obtained from 28 patients who underwent the operation of lung cancer in the Department of Surgery, Hamamatsu University Hospital. This provided us with 13 cases of adenocarcinoma and 15 cases of squamous cell carcinoma histologically. Tumor tissue and adjacent normal lung tissue were excised and washed well by saline. All samples were quickly frozen and stored at -80°C until homogenization. All samples were homogenized with 10 times as much volume of buffer (0.1M Tris, 0.15M NaCl pH7.5) by microhomogenizer Physcotron(NINI-ON Tokyo Japan). After keeping in ice-water for 3 hours, homogenized samples were centrifuged at 16,000 X G at 4°C for 10 min. Supernatants were stored at-80°C until assayed.

Enzyme immunoassay
T-PA, UK, and PAI-1 antigens were measured by enzyme immunoassay, previously reported from our laboratory.[6-8] PAI-2 antigen levels were measured by ELISA emplying assay kit (TintElize PAI-2, Biopool, Sweden) .

Protein concentration

The protein concentration of homogenized samples was measured by BCA Protein Assay Reagant (Pierce, U.S.A.)

Immunohistochemistry

Localization of t-PA, UK, PAI-1, and PAI-2 in cancer tissue was determined by immunohistochemistry. Staining was performed by ABC method (VECTASTAIN ABC KIT, VECTOR LABORATORIES, U.S.A.). Monoclonal antibodies to human UK(#394 American Diagnosyica Inc., U.S.A.), PAI-1(MAI-11,Biopool, Sweden), and PAI-2 (MAI-21, Biopool,Sweden) were used. Counterstaining was done by methyl green.

Statistics

For comparison between groups, Student's t-test and Pearsons' correlation coefficient were used. Differences were considered as significanct below P value=0.05

RESULTS

Concentration of UK, t-PA, PAI-1,t-PA·PAI-1 complex and PAI-2 in normal and cancer tissue

Table 1 shows the values of UK, t-PA, PAI-1, t-PA·PAI-1 complex, and PAI-2 levels in cancer tissues and normal lung tissues. UK and PAI-2 levels in cancer tissues were significantly higher than those in normal tissues. T-PA levels in cancer tissues were significantly lower than those in normal tissues. Total and free PAI-1 were significantly higher in cancer tissues than those in normal tissues. We compared these antigen levels between adenocarcinoma and squamous cell carcinoma. There were no significant differences between two types of carcinoma.(Data not shown)

Table 1   Concentrations of UK, t-PA, PAI-1, t-PA·PAI-1 complex and PAI-2 in tissues.

	Normal tissue (n=22)	Cancer tissue (n=28)
	(ng/mg protein)	(ng/mg protein)
UK	1.70±0.27	4.18±0.65*
t-PA	24.01±6.40	6.99±1.26*
PAI-1(t)	1.66±1.17	3.35±0.37*
PAI-1(c)	1.45±0.26	2.09±0.24
PAI-1(f)	0.20±0.08	1.26±0.30*
PAI-2	0.97±0.31	7.67±1.86*

The mean value±SE are shown.   *P<0.05
PAI-1(t): total PAI-1, PA1-1(c): t-PA·PAI-1 complex
PAI-1(f): free PAI-1

<u>Relationship between tissue levels of UK, t-PA, PAI-1,t-PA·PAI-1 complex and PAI-2, and maximal diameter of tumor.</u>

We examined relationship between tissue levels of these parameters and maximal diameter of cancer tissue to compare the difference of the value between early stage and progressed stage.   We devided tumor specimens into two groups by the diameter of a tumor.   In one group, the maximal diameter of cancer tissue was 3 centimeters or less, and in another group, the diameter was more than 3 centimeters.   In T1 stage, the maximal diameter of cancer tissue was 3 centimeters or less, which usually indicates early clinical stage on the basis of International Staging System for Lung Cancer.   The data was shown in Table 2.   The levels of UK, PAI-1, t-PA·PAI-1complex and PAI-2 antigens had tendency to increase in a group with larger size.   On the contrary, the level of t-PA antigen had tendency to decrease in a group with larger size.

Table 2   Relationship between tissue levels of UK, t-PA, PAI-1,
t-PA·PAI-1complex and PAI-2,   and maximal diameter of cancer tissue.

	M.D (≤3cm) (n=9)	M.D.(3cm<) (n=19)
	(ng/mg protein)	(ng/mg protein)
UK	2.87±0.68	4.60±0.80
t-PA	8.10±3.10	6.34±1.16
PAI-1(t)	2.46±0.42	3.70±0.47
PAI-1(c)	1.68±0.26	2.25±0.32
PAI-1(f)	0.78±0.28	1.46±0.40
PAI-2	5.65±1.98	8.48±1.45

The mean value±SE are shown.
PAI-1(t): total PAI-1, PAI-1(c): t-PA·PAI-1 complex
PAI-1(f): free PAI-1   M.D. : maximal diameter of cancer tissue

<u>Relationship between tissue levels of UK, t-PA, PAI-1, t-PA·PAI-1 complex and PAI-2, and lymph node involvements.</u>

We examined whether the lymph node involvements were related to the concentration of UK, t-PA, PAI-1, t-PA·PAI-1complex and PAI-2 antigens. Two groups were classified as those without lymph node involvements, and those with metastatic lymph node involvements in hilum and mediastinum.   There was no significant difference in the levels of UK, t-PA, PAI-1 and t-PA·PAI-1 complex between two groups.   The levels of PAI-2 antigens in case with lymph node involvements were significantly lower than those without lymph node involvements.(Table 3)

210

Table 3   Relationship between tissue levels of UK, t-PA, PAI-1,
t-PA•PAI-1 complex and PAI-2, and lymph node involvements.

	LN involvements(-) (n=13)	LN involvements(+) (n=15)
	(ng/mg protein)	(ng/mg protein)
UK	5.09±1.04	3.22±0.61
t-PA	7.32±1.81	6.49±1.64
PAI-1(t)	3.37±0.37	3.33±0.63
PAI-1(c)	2.27±1.10	1.92±0.36
PAI-1(f)	1.09±0.31	1.41±0.50
PAI-2	12.01±3.29	3.92±1.30*

The mean value±SE are shown.   *P<0.05
PA1-1(t): total PAI-1, PAI-1(c): t-PA•PAI-1 complex
PAI-1(f): free PAI-1   LN: lymph node

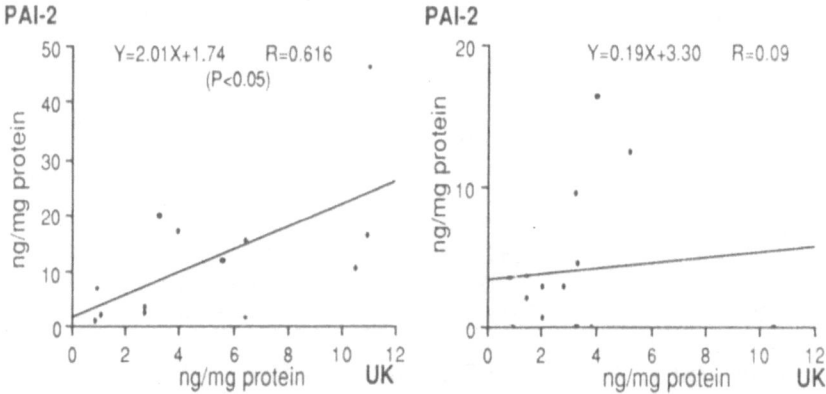

Figure.1   Correlation between PAI-2 level and UK level in the absence(left)
and presence(right) of lymphnode involvements.

Next we examined the relationship between UK, and PAI-1 and PAI-2 as to
lymph node involvements. In the absence of lymph node involvements, PAI-2
antigen had significantly positive correlation with UK antigen, however, in
the presence of lymph node involvements, there was no significant
correlation between PAI-2 antigen level and UK antigen level.(Fig.1)
On the contrary, in the absence or presence of lymph node involvements,
there was no significant correlation between PAI-1 antigen level and UK
antigen level. (Data not shown)

Immunohistological staining
Figure 2 shows stainig of cancer tissues by immunohistological methods.

UK, PAI-1, and PAI-2 were stained in a cytosol of cancer cells, but not in adjacent healthy normal tissues. The staining of UK, PAI-1, and PAI-2 antigen showed close similarity in localization within cancer tissues.

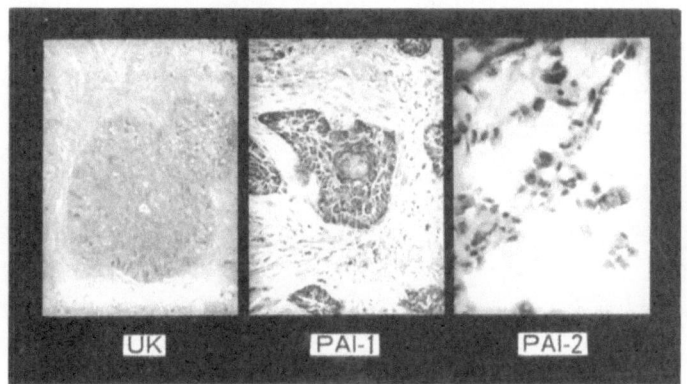

Figure.2   Staining of UK, PAI-1, and PAI-2 in cancer tissues

DISCUSSION

In the study, we examined the concentrations of UK, t-PA, PAI-1 and PAI-2 in normal lung tissues and cancer tissues, and investigated whether these levels are related to the extent of progression of the tumor and lymph node involvements.

UK antigen levels in cancer tissues were significantly higher than those in normal tissues, however there was no significant difference between tissue types of cancer, and no significant correlation with tumor size and lymph node involvements.   Many reports in various kinds of cancer have been reported in this regards, although, there has been controversy.[9-10]

It has been reported that t-PA antigen levels are lower in cancer tissues.[11] This agrees with our results.   The reason why decrease in t-PA antigen levels in cancer tissues might be a lesser vascularization related with cancer invasion

The levels of PAI-1 antigen in cancer tissues were higher than those in normal tissues.   However, the major part of increased PAI-1 was free PAI-1, and there was no significant correlation between PAI-1 antigen levels and UK antigen levels.   These results suggest that PAI-1 did not increase in response to increase in UK level, and it may deny the possibility that PAI-1 plays an important role to prevent tumor growth and invasion by inhibiting UK activity.[12]   Further investigation should be required in order to find exact relationship between UK and PAI-1.

Recently it has been reported that PAI-2 specifically inhibits UK and even receptor-bound UK in human blood monocytes and colon cancer cell line.[13-14] In our study, PAI-2 antigen had significant positive correlation with UK antigen in the absence of lymph node involvements and there was no